# *A Handbook of* **Statistical Analys** *Using* **R**

# *A Handbook of* **Statistical Analyses** *Using* **R**

Brian S. Everitt
Torsten Hothorn

Published in 2006 by
Chapman & Hall/CRC
Taylor & Francis Group
6000 Broken Sound Parkway NW, Suite 300
Boca Raton, FL 33487-2742


Printed in the United States of America on acid-free paper
10 9 8 7 6 5

International Standard Book Number-10: 1-58488-539-4 (Softcover)
International Standard Book Number-13: 978-1-58488-539-9 (Softcover)



**Library of Congress Cataloging-in-Publication Data**

Catalog record is available from the Library of Congress

Taylor & Francis Group
is the Academic Division of Informa plc.

**Visit the Taylor & Francis Web site at**
**http://www.taylorandfrancis.com**

**and the CRC Press Web site at**
**http://www.crcpress.com**

Dedication

To the families of both authors
for their constant support and encouragement

# Preface

This book is intended as a guide to data analysis with the R system for statistical computing. R is an environment incorporating an implementation of the S programming language, which is powerful, flexible and has excellent graphical facilities (R Development Core Team, 2005b). In the Handbook we aim to give relatively brief and straightforward descriptions of how to conduct a range of statistical analyses using R. Each chapter deals with the analysis appropriate for one or several data sets. A brief account of the relevant statistical background is included in each chapter along with appropriate references, but our prime focus is on how to use R and how to interpret results. We hope the book will provide students and researchers in many disciplines with a self-contained means of using R to analyse their data.

R is an open-source project developed by dozens of volunteers for more than ten years now and is available from the Internet under the General Public Licence. R has become the *lingua franca* of statistical computing. Increasingly, implementations of new statistical methodology first appear as R add-on packages. In some communities, such as in bioinformatics, R already is the primary workhorse for statistical analyses. Because the sources of the R system are open and available to everyone without restrictions and because of its powerful language and graphical capabilities, R has started to become the main computing engine for reproducible statistical research (Leisch, 2002a,b, 2003, Leisch and Rossini, 2003, Gentleman, 2005). For a reproducible piece of research, the original observations, all data preprocessing steps, the statistical analysis as well as the scientific report form a unity and all need to be available for inspection, reproduction and modification by the readers.

Reproducibility is a natural requirement for textbooks such as the 'Handbook of Statistical Analyses Using R' and therefore this book is fully reproducible using an R version greater or equal to 2.2.1. All analyses and results, including figures and tables, can be reproduced by the reader without having to retype a single line of R code. The data sets presented in this book are collected in a dedicated add-on package called *HSAUR* accompanying this book. The package can be installed from the Comprehensive R Archive Network (CRAN) via

```
R> install.packages("HSAUR")
```

and its functionality is attached by

```
R> library("HSAUR")
```

The relevant parts of each chapter are available as a *vignette*, basically a document including both the R sources and the rendered output of every analysis contained in the book. For example, the first chapter can be inspected by

```
R> vignette("Ch_introduction_to_R", package = "HSAUR")
```

and the R sources are available for reproducing our analyses by

```
R> edit(vignette("Ch_introduction_to_R", package = "HSAUR"))
```

An overview on all chapter vignettes included in the package can be obtained from

```
R> vignette(package = "HSAUR")
```

We welcome comments on the R package *HSAUR*, and where we think these add to or improve our analysis of a data set we will incorporate them into the package and, hopefully at a later stage, into a revised or second edition of the book.

Plots and tables of results obtained from R are all labelled as 'Figures' in the text. For the graphical material, the corresponding figure also contains the 'essence' of the R code used to produce the figure, although this code may differ a little from that given in the *HSAUR* package, since the latter may include some features, for example thicker line widths, designed to make a basic plot more suitable for publication.

We would like to thank the R Development Core Team for the R system, and authors of contributed add-on packages, particularly Uwe Ligges and Vince Carey for helpful advice on *scatterplot3d* and *gee*. Kurt Hornik, Ludwig A. Hothorn, Fritz Leisch and Rafael Weißbach provided good advice with some statistical and technical problems. We are also very grateful to Achim Zeileis for reading the entire manuscript, pointing out inconsistencies or even bugs and for making many suggestions which have led to improvements. Lastly we would like to thank the CRC Press staff, in particular Rob Calver, for their support during the preparation of the book. Any errors in the book are, of course, the joint responsibility of the two authors.

**Brian S. Everitt and Torsten Hothorn**
London and Erlangen, December 2005

# List of Figures

# List of Tables

# Contents

CHAPTER 1

# An Introduction to R

## 1.1 What Is R?

The R system for statistical computing is an environment for data analysis and graphics. The root of R is the S language, developed by John Chambers and colleagues (Becker et al., 1988, Chambers and Hastie, 1992, Chambers, 1998) at Bell Laboratories (formerly AT&T, now owned by Lucent Technologies) starting in the 1960s. The S language was designed and developed as a programming language for data analysis tasks but in fact it is a full-featured programming language in its current implementations.

The development of the R system for statistical computing is heavily influenced by the open source idea: The base distribution of R and a large number of user contributed extensions are available under the terms of the Free Software Foundation's GNU General Public License in source code form. This licence has two major implications for the data analyst working with R. The complete source code is available and thus the practitioner can investigate the details of the implementation of a special method, can make changes and can distribute modifications to colleagues. As a side-effect, the R system for statistical computing is available to everyone. All scientists, especially including those working in developing countries, have access to state-of-the-art tools for statistical data analysis without additional costs. With the help of the R system for statistical computing, research really becomes reproducible when both the data and the results of all data analysis steps reported in a paper are available to the readers through an R transcript file. R is most widely used for teaching undergraduate and graduate statistics classes at universities all over the world because students can freely use the statistical computing tools.

The base distribution of R is maintained by a small group of statisticians, the R Development Core Team. A huge amount of additional functionality is implemented in add-on packages authored and maintained by a large group of volunteers. The main source of information about the R system is the world wide web with the official home page of the R project being

`http://www.R-project.org`

All resources are available from this page: the R system itself, a collection of add-on packages, manuals, documentation and more.

The intention of this chapter is to give a rather informal introduction to basic concepts and data manipulation techniques for the R novice. Instead of a rigid treatment of the technical background, the most common tasks are illustrated by practical examples and it is our hope that this will enable readers to get started without too many problems.

## 1.2 Installing R

The R system for statistical computing consists of two major parts: the base system and a collection of user contributed add-on packages. The R language is implemented in the base system. Implementations of statistical and graphical procedures are separated from the base system and are organised in the form of packages. A package is a collection of functions, examples and documentation. The functionality of a package is often focused on a special statistical methodology. Both the base system and packages are distributed via the Comprehensive R Archive Network (CRAN) accessible under

`http://CRAN.R-project.org`

### *1.2.1 The Base System and the First Steps*

The base system is available in source form and in precompiled form for various Unix systems, Windows platforms and Mac OS X. For the data analyst, it is sufficient to download the precompiled binary distribution and install it locally. Windows users follow the link

`http://CRAN.R-project.org/bin/windows/base/release.htm`

download the corresponding file (currently named `rw2021.exe`), execute it locally and follow the instructions given by the installer.

Depending on the operating system, R can be started either by typing '`R`' on the shell (Unix systems) or by clicking on the R symbol (as shown left) created by the installer (Windows).

R comes without any frills and on start up shows simply a short introductory message including the version number and a prompt '`>`':

```
R : Copyright 2005 The R Foundation for Statistical Computing
Version 2.2.1 (2005-12-20), ISBN 3-900051-07-0

R is free software and comes with ABSOLUTELY NO WARRANTY.
You are welcome to redistribute it under certain conditions.
Type 'license()' or 'licence()' for distribution details.

R is a collaborative project with many contributors.
Type 'contributors()' for more information and
'citation()' on how to cite R or R packages in publications.

Type 'demo()' for some demos, 'help()' for on-line help, or
'help.start()' for an HTML browser interface to help.
Type 'q()' to quit R.

>
```

One can change the appearance of the prompt by

```
> options(prompt = "R> ")
```

and we will use the prompt R> for the display of the code examples throughout this book.

Essentially, the R system evaluates commands typed on the R prompt and returns the results of the computations. The end of a command is indicated by the return key. Virtually all introductory texts on R start with an example using R as pocket calculator, and so do we:

```
R> x <- sqrt(25) + 2
```

This simple statement asks the R interpreter to calculate $\sqrt{25}$ and then to add 2. The result of the operation is assigned to an R object with variable name `x`. The assignment operator `<-` binds the value of its right hand side to a variable name on the left hand side. The value of the object `x` can be inspected simply by typing

```
R> x
[1] 7
```

which, implicitly, calls the `print` method:

```
R> print(x)
[1] 7
```

### *1.2.2 Packages*

The base distribution already comes with some high-priority add-on packages, namely

| | | | |
|---|---|---|---|
| boot | nlme | KernSmooth | MASS |
| base | class | cluster | datasets |
| foreign | grDevices | graphics | grid |
| lattice | methods | mgcv | nnet |
| rpart | spatial | splines | stats |
| stats4 | survival | tcltk | tools |
| utils | | | |

The packages listed here implement standard statistical functionality, for example linear models, classical tests, a huge collection of high-level plotting functions or tools for survival analysis; many of these will be described and used in later chapters.

Packages not included in the base distribution can be installed directly from the R prompt. At the time of writing this chapter, 640 user contributed packages covering almost all fields of statistical methodology were available.

Given that an Internet connection is available, a package is installed by supplying the name of the package to the function `install.packages`. If, for example, add-on functionality for robust estimation of covariance matrices via sandwich estimators is required (for example in Chapter 11), the *sandwich* package (Zeileis, 2004) can be downloaded and installed via

```
R> install.packages("sandwich")
```

The package functionality is available after *attaching* the package by

```
R> library("sandwich")
```

A comprehensive list of available packages can be obtained from

`http://CRAN.R-project.org/src/contrib/PACKAGES.html`

Note that on Windows operating systems, precompiled versions of packages are downloaded and installed. In contrast, packages are compiled locally before they are installed on Unix systems.

## 1.3 Help and Documentation

Roughly, three different forms of documentation for the R system for statistical computing may be distinguished: online help that comes with the base distribution or packages, electronic manuals and publications work in the form of books etc.

The help system is a collection of manual pages describing each user-visible function and data set that comes with R. A manual page is shown in a pager or web browser when the name of the function we would like to get help for is supplied to the `help` function

```
R> help("mean")
```

or, for short,

```
R> ?mean
```

Each manual page consists of a general description, the argument list of the documented function with a description of each single argument, information about the return value of the function and, optionally, references, cross-links and, in most cases, executable examples. The function `help.search` is helpful for searching within manual pages. An overview on documented topics in an add-on package is given, for example for the *sandwich* package, by

```
R> help(package = "sandwich")
```

Often a package comes along with an additional document describing the package functionality and giving examples. Such a document is called a *vignette* (Leisch, 2003, Gentleman, 2005). The *sandwich* package vignette is opened using

```
R> vignette("sandwich")
```

More extensive documentation is available electronically from the collection of manuals at

`http://CRAN.R-project.org/manuals.html`

For the beginner, at least the first and the second document of the following four manuals (R Development Core Team, 2005a,c,d,e) are mandatory:

**An Introduction to R:** A more formal introduction to data analysis with R than this chapter.

**R Data Import/Export:** A very useful description of how to read and write various external data formats.

**R Installation and Administration:** Hints for installing R on special platforms.

**Writing R Extensions:** The authoritative source on how to write R programs and packages.

Both printed and online publications are available, the most important ones are 'Modern Applied Statistics with S' (Venables and Ripley, 2002), 'Introductory Statistics with R' (Dalgaard, 2002), 'R Graphics' (Murrell, 2005) and the R Newsletter, freely available from

`http://CRAN.R-project.org/doc/Rnews/`

In case the electronically available documentation and the answers to frequently asked questions (FAQ), available from

`http://CRAN.R-project.org/faqs.html`

have been consulted but a problem or question remains unsolved, the `r-help` email list is the right place to get answers to well-thought-out questions. It is helpful to read the posting guide

`http://www.R-project.org/posting-guide.html`

before starting to ask.

## 1.4 Data Objects in R

The data handling and manipulation techniques explained in this chapter will be illustrated by means of a data set of 2000 world leading companies, the Forbes 2000 list for the year 2004 collected by 'Forbes Magazine'. This list is originally available from

`http://www.forbes.com`

and, as an R data object, it is part of the *HSAUR* package (*Source*: From Forbes.com, New York, 2004. With permission.). In a first step, we make the data available for computations within R. The `data` function searches for data objects of the specified name (`"Forbes2000"`) in the package specified via the `package` argument and, if the search was successful, attaches the data object to the global environment:

```
R> data("Forbes2000", package = "HSAUR")
R> ls()
```

```
[1] "Forbes2000" "x"
```

The output of the `ls` function lists the names of all objects currently stored in the global environment, and, as the result of the previous command, a variable named `Forbes2000` is available for further manipulation. The variable `x` arises from the pocket calculator example in Subsection 1.2.1.

As one can imagine, printing a list of 2000 companies via

```
R> print(Forbes2000)
```

```
  rank                 name       country      category  sales
1    1            Citigroup United States       Banking  94.71
2    2     General Electric United States Conglomerates 134.19
3    3 American Intl Group United States     Insurance  76.66
  profits  assets marketvalue
1   17.85 1264.03      255.30
2   15.59  626.93      328.54
3    6.46  647.66      194.87
...
```

will not be particularly helpful in gathering some initial information about the data; it is more useful to look at a description of their structure found by using the following command

```
R> str(Forbes2000)
```

```
'data.frame':         2000 obs. of  8 variables:
 $ rank        : int  1 2 3 4 5 ...
 $ name        : chr  "Citigroup" "General Electric" ...
 $ country     : Factor w/ 61 levels "Africa","Australia",..:
    60 60 60 60 56 ...
 $ category    : Factor w/ 27 levels "Aerospace & defense",..:
    2 6 16 19 19 ...
 $ sales       : num   94.7 134.2 ...
 $ profits     : num  17.9 15.6 ...
 $ assets      : num  1264  627 ...
 $ marketvalue: num  255 329 ...
```

The output of the `str` function tells us that `Forbes2000` is an object of class *data.frame*, the most important data structure for handling tabular statistical data in R. As expected, information about 2000 observations, i.e., companies, are stored in this object. For each observation, the following eight variables are available:

`rank`: the ranking of the company,

`name`: the name of the company,

`country`: the country the company is situated in,

`category`: a category describing the products the company produces,

`sales`: the amount of sales of the company in billion US dollars,

`profits`: the profit of the company in billion US dollars,

`assets`: the assets of the company in billion US dollars,

`marketvalue`: the market value of the company in billion US dollars.

A similar but more detailed description is available from the help page for the `Forbes2000` object:

```
R> help("Forbes2000")
```

or

```
R> ?Forbes2000
```

All information provided by `str` can be obtained by specialised functions as well and we will now have a closer look at the most important of these.

The R language is an object-oriented programming language, so every object is an instance of a class. The name of the class of an object can be determined by

```
R> class(Forbes2000)
[1] "data.frame"
```

Objects of class *data.frame* represent data the traditional table oriented way. Each row is associated with one single observation and each column corresponds to one variable. The dimensions of such a table can be extracted using the `dim` function

```
R> dim(Forbes2000)
[1] 2000    8
```

Alternatively, the numbers of rows and columns can be found using

```
R> nrow(Forbes2000)
[1] 2000
R> ncol(Forbes2000)
[1] 8
```

The results of both statements show that `Forbes2000` has 2000 rows, i.e., observations, the companies in our case, with eight variables describing the observations. The variable names are accessible from

```
R> names(Forbes2000)
[1] "rank"          "name"          "country"       "category"
[5] "sales"         "profits"       "assets"        "marketvalue"
```

The values of single variables can be extracted from the `Forbes2000` object by their names, for example the ranking of the companies

```
R> class(Forbes2000[, "rank"])
[1] "integer"
```

is stored as an integer variable. Brackets `[]` always indicate a subset of a larger object, in our case a single variable extracted from the whole table. Because *data.frames* have two dimensions, observations and variables, the comma is required in order to specify that we want a subset of the second dimension, i.e., the variables. The rankings for all 2000 companies are represented in a *vector* structure the length of which is given by

```
R> length(Forbes2000[, "rank"])
[1] 2000
```

A *vector* is the elementary structure for data handling in R and is a set of simple elements, all being objects of the same class. For example, a simple vector of the numbers one to three can be constructed by one of the following commands

```
R> 1:3
[1] 1 2 3
R> c(1, 2, 3)
[1] 1 2 3
R> seq(from = 1, to = 3, by = 1)
[1] 1 2 3
```

The unique names of all 2000 companies are stored in a character vector

```
R> class(Forbes2000[, "name"])
[1] "character"
R> length(Forbes2000[, "name"])
[1] 2000
```

and the first element of this vector is

```
R> Forbes2000[, "name"][1]
[1] "Citigroup"
```

Because the companies are ranked, Citigroup is the world's largest company according to the Forbes 2000 list. Further details on vectors and subsetting are given in Section 1.6.

Nominal measurements are represented by *factor* variables in R, such as the category of the company's business segment

```
R> class(Forbes2000[, "category"])
[1] "factor"
```

Objects of class *factor* and *character* basically differ in the way their values are stored internally. Each element of a vector of class *character* is stored as a *character* variable whereas an integer variable indicating the level of a *factor* is saved for *factor* objects. In our case, there are

```
R> nlevels(Forbes2000[, "category"])
[1] 27
```

different levels, i.e., business categories, which can be extracted by

```
R> levels(Forbes2000[, "category"])
[1] "Aerospace & defense"
[2] "Banking"
[3] "Business services & supplies"
...
```

As a simple summary statistic, the frequencies of the levels of such a *factor* variable can be found from

```
R> table(Forbes2000[, "category"])
```

```
         Aerospace & defense                                   Banking
                          19                                       313
Business services & supplies
                          70
...
```

The sales, assets, profits and market value variables are of type `numeric`, the natural data type for continuous or discrete measurements, for example

```
R> class(Forbes2000[, "sales"])
[1] "numeric"
```

and simple summary statistics such as the mean, median and range can be found from

```
R> median(Forbes2000[, "sales"])
[1] 4.365
R> mean(Forbes2000[, "sales"])
[1] 9.69701
R> range(Forbes2000[, "sales"])
[1]   0.01 256.33
```

The `summary` method can be applied to a numeric vector to give a set of useful summary statistics namely the minimum, maximum, mean, median and the 25% and 75% quartiles; for example

```
R> summary(Forbes2000[, "sales"])
   Min. 1st Qu.  Median    Mean 3rd Qu.    Max.
  0.010   2.018   4.365   9.697   9.547 256.300
```

## 1.5 Data Import and Export

In the previous section, the data from the Forbes 2000 list of the world's largest companies were loaded into R from the *HSAUR* package but we will now explore practically more relevant ways to import data into the R system. The most frequent data formats the data analyst is confronted with are comma separated files, Excel spreadsheets, files in SPSS format and a variety of SQL data base engines. Querying data bases is a non-trivial task and requires additional knowledge about querying languages and we therefore refer to the 'R Data Import/Export' manual – see Section 1.3. We assume that a comma separated file containing the Forbes 2000 list is available as `Forbes2000.csv` (such a file is part of the *HSAUR* source package in directory `HSAUR/inst/rawdata`). When the fields are separated by commas and each row begins with a name (a text format typically created by Excel), we can read in the data as follows using the `read.table` function

```
R> csvForbes2000 <- read.table("Forbes2000.csv", header = TRUE,
+      sep = ",", row.names = 1)
```

The argument `header = TRUE` indicates that the entries in the first line of the text file `"Forbes2000.csv"` should be interpreted as variable names. Columns are separated by a comma (`sep = ","`), users of continental versions of Excel should take care of the character symbol coding for decimal points (by default `dec = "."`). Finally, the first column should be interpreted as row names but not as a variable (`row.names = 1`). Alternatively, the function `read.csv` can be used to read comma separated files. The function `read.table` by default guesses the class of each variable from the specified file. In our case, character variables are stored as factors

```
R> class(csvForbes2000[, "name"])
[1] "factor"
```

which is only suboptimal since the names of the companies are unique. However, we can supply the types for each variable to the `colClasses` argument

```
R> csvForbes2000 <- read.table("Forbes2000.csv", header = TRUE,
+     sep = ",", row.names = 1, colClasses = c("character",
+         "integer", "character", "factor", "factor",
+         "numeric", "numeric", "numeric", "numeric"))
R> class(csvForbes2000[, "name"])
[1] "character"
```

and check if this object is identical with our previous Forbes 2000 list object

```
R> all.equal(csvForbes2000, Forbes2000)
[1] TRUE
```

The argument `colClasses` expects a character vector of length equal to the number of columns in the file. Such a vector can be supplied by the `c` function that combines the objects given in the parameter list into a *vector*

```
R> classes <- c("character", "integer", "character",
+     "factor", "factor", "numeric", "numeric", "numeric",
+     "numeric")
R> length(classes)
[1] 9
R> class(classes)
[1] "character"
```

An R interface to the open data base connectivity standard (ODBC) is available in package *RODBC* and its functionality can be used to assess Excel and Access files directly:

```
R> library("RODBC")
R> cnct <- odbcConnectExcel("Forbes2000.xls")
R> sqlQuery(cnct, "select * from \"Forbes2000$\"")
```

The function `odbcConnectExcel` opens a connection to the specified Excel or Access file which can be used to send SQL queries to the data base engine and retrieve the results of the query.

Files in SPSS format are read in a way similar to reading comma separated files, using the function `read.spss` from package *foreign* (which comes with the base distribution).

Exporting data from R is now rather straightforward. A comma separated file readable by Excel can be constructed from a *data.frame* object via

```
R> write.table(Forbes2000, file = "Forbes2000.csv",
+      sep = ",", col.names = NA)
```

The function `write.csv` is one alternative and the functionality implemented in the *RODBC* package can be used to write data directly into Excel spreadsheets as well.

Alternatively, when data should be saved for later processing in R only, R objects of arbitrary kind can be stored into an external binary file via

```
R> save(Forbes2000, file = "Forbes2000.rda")
```

where the extension `.rda` is standard. We can get the file names of all files with extension `.rda` from the working directory

```
R> list.files(pattern = "\\.rda")
```

```
[1] "Forbes2000.rda"
```

and we can load the contents of the file into R by

```
R> load("Forbes2000.rda")
```

## 1.6 Basic Data Manipulation

The examples shown in the previous section have illustrated the importance of *data.frames* for storing and handling tabular data in R. Internally, a *data.frame* is a *list* of vectors of a common length $n$, the number of rows of the table. Each of those vectors represents the measurements of one variable and we have seen that we can access such a variable by its name, for example the names of the companies

```
R> companies <- Forbes2000[, "name"]
```

Of course, the `companies` vector is of class *character* and of length 2000. A subset of the elements of the vector `companies` can be extracted using the `[]` subset operator. For example, the largest of the 2000 companies listed in the Forbes 2000 list is

```
R> companies[1]
```

```
[1] "Citigroup"
```

and the top three companies can be extracted utilising an integer vector of the numbers one to three:

```
R> 1:3
```

```
[1] 1 2 3
```

```
R> companies[1:3]
```

```
[1] "Citigroup"             "General Electric"
[3] "American Intl Group"
```

In contrast to indexing with positive integers, negative indexing returns all elements which are *not* part of the index vector given in brackets. For example, all companies except those with numbers four to two-thousand, i.e., the top three companies, are again

```
R> companies[-(4:2000)]
```

```
[1] "Citigroup"             "General Electric"
[3] "American Intl Group"
```

The complete information about the top three companies can be printed in a similar way. Because *data.frames* have a concept of rows and columns, we need to separate the subsets corresponding to rows and columns by a comma. The statement

```
R> Forbes2000[1:3, c("name", "sales", "profits", "assets")]
```

```
                 name  sales profits  assets
1           Citigroup  94.71   17.85 1264.03
2    General Electric 134.19   15.59  626.93
3 American Intl Group  76.66    6.46  647.66
```

extracts the variables `name`, `sales`, `profits` and `assets` for the three largest companies. Alternatively, a single variable can be extracted from a *data.frame* by

```
R> companies <- Forbes2000$name
```

which is equivalent to the previously shown statement

```
R> companies <- Forbes2000[, "name"]
```

We might be interested in extracting the largest companies with respect to an alternative ordering. The three top selling companies can be computed along the following lines. First, we need to compute the ordering of the companies' sales

```
R> order_sales <- order(Forbes2000$sales)
```

which returns the indices of the ordered elements of the numeric vector `sales`. Consequently the three companies with the lowest sales are

```
R> companies[order_sales[1:3]]
```

```
[1] "Custodia Holding"         "Central European Media"
[3] "Minara Resources"
```

The indices of the three top sellers are the elements 1998, 1999 and 2000 of the integer vector `order_sales`

```
R> Forbes2000[order_sales[c(2000, 1999, 1998)], c("name",
+     "sales", "profits", "assets")]
```

```
              name  sales profits assets
10 Wal-Mart Stores 256.33    9.05 104.91
5               BP 232.57   10.27 177.57
4       ExxonMobil 222.88   20.96 166.99
```

Another way of selecting vector elements is the use of a logical vector being TRUE when the corresponding element is to be selected and FALSE otherwise. The companies with assets of more than 1000 billion US dollars are

```
R> Forbes2000[Forbes2000$assets > 1000, c("name", "sales",
+     "profits", "assets")]
```

```
                 name sales profits  assets
1           Citigroup 94.71   17.85 1264.03
9          Fannie Mae 53.13    6.48 1019.17
403 Mizuho Financial 24.40  -20.11 1115.90
```

where the expression Forbes2000$assets > 1000 indicates a logical vector of length 2000 with

```
R> table(Forbes2000$assets > 1000)
```

```
FALSE  TRUE
 1997     3
```

elements being either FALSE or TRUE. In fact, for some of the companies the measurement of the profits variable are missing. In R, missing values are treated by a special symbol, NA, indicating that this measurement is not available. The observations with profit information missing can be obtained via

```
R> na_profits <- is.na(Forbes2000$profits)
R> table(na_profits)
```

```
na_profits
FALSE  TRUE
 1995     5
```

```
R> Forbes2000[na_profits, c("name", "sales", "profits",
+     "assets")]
```

```
                        name sales profits assets
772                      AMP  5.40      NA  42.94
1085                     HHG  5.68      NA  51.65
1091                     NTL  3.50      NA  10.59
1425       US Airways Group  5.50      NA   8.58
1909 Laidlaw International  4.48      NA   3.98
```

where the function is.na returns a logical vector being TRUE when the corresponding element of the supplied vector is NA. A more comfortable approach is available when we want to remove all observations with at least one missing value from a *data.frame* object. The function complete.cases takes a *data.frame* and returns a logical vector being TRUE when the corresponding observation does not contain any missing value:

```
R> table(complete.cases(Forbes2000))
```

```
FALSE  TRUE
    5  1995
```

Subsetting *data.frames* driven by logical expressions may induce a lot of typing which can be avoided. The subset function takes a *data.frame* as first argument and a logical expression as second argument. For example, we can select a subset of the Forbes 2000 list consisting of all companies situated in the United Kingdom by

```
R> UKcomp <- subset(Forbes2000, country == "United Kingdom")
R> dim(UKcomp)
```

```
[1] 137   8
```

i.e., 137 of the 2000 companies are from the UK. Note that it is not necessary to extract the variable `country` from the *data.frame* `Forbes2000` when formulating the logical expression.

## 1.7 Simple Summary Statistics

Two functions are helpful for getting an overview about R objects: `str` and `summary`, where `str` is more detailed about data types and `summary` gives a collection of sensible summary statistics. For example, applying the `summary` method to the `Forbes2000` data set,

```
R> summary(Forbes2000)
```

results in the following output

```
      rank              name                        country   
 Min.   :   1.0   Length:2000          United States :751  
 1st Qu.: 500.8   Class :character     Japan         :316  
 Median :1000.5   Mode  :character     United Kingdom:137  
 Mean   :1000.5                        Germany       : 65  
 3rd Qu.:1500.2                        France        : 63  
 Max.   :2000.0                        Canada        : 56  
                                        (Other)       :612  
                       category          sales        
 Banking                   : 313   Min.   :  0.010  
 Diversified financials: 158   1st Qu.:  2.018  
 Insurance                 : 112   Median :  4.365  
 Utilities                 : 110   Mean   :  9.697  
 Materials                 :  97   3rd Qu.:  9.547  
 Oil & gas operations      :  90   Max.   :256.330  
 (Other)                   :1120                    
    profits              assets             marketvalue    
 Min.   :-25.8300   Min.   :   0.270   Min.   :  0.02  
 1st Qu.:  0.0800   1st Qu.:   4.025   1st Qu.:  2.72  
 Median :  0.2000   Median :   9.345   Median :  5.15  
 Mean   :  0.3811   Mean   :  34.042   Mean   : 11.88  
 3rd Qu.:  0.4400   3rd Qu.:  22.793   3rd Qu.: 10.60  
 Max.   : 20.9600   Max.   :1264.030   Max.   :328.54  
 NA's   :  5.0000                                      
```

From this output we can immediately see that most of the companies are situated in the US and that most of the companies are working in the banking sector as well as that negative profits, or losses, up to 26 billion US dollars occur.

Internally, summary is a so-called *generic function* with methods for a multitude of classes, i.e., summary can be applied to objects of different classes and will report sensible results. Here, we supply a *data.frame* object to summary where it is natural to apply summary to each of the variables in this *data.frame*. Because a *data.frame* is a *list* with each variable being an element of that *list*, the same effect can be achieved by

```
R> lapply(Forbes2000, summary)
```

The members of the apply family help to solve recurring tasks for each element of a *data.frame*, *matrix*, *list* or for each level of a factor. It might be interesting to compare the profits in each of the 27 categories. To do so, we first compute the median profit for each category from

```
R> mprofits <- tapply(Forbes2000$profits, Forbes2000$category,
+      median, na.rm = TRUE)
```

a command that should be read as follows. For each level of the factor category, determine the corresponding elements of the numeric vector profits and supply them to the median function with additional argument na.rm = TRUE. The latter one is necessary because profits contains missing values which would lead to a non-sensible result of the median function

```
R> median(Forbes2000$profits)
```

```
[1] NA
```

The three categories with highest median profit are computed from the vector of sorted median profits

```
R> rev(sort(mprofits))[1:3]
```

```
         Oil & gas operations           Drugs & biotechnology
                         0.35                            0.35
Household & personal products
                         0.31
```

where rev rearranges the vector of median profits sorted from smallest to largest. Of course, we can replace the median function with mean or whatever is appropriate in the call to tapply. In our situation, mean is not a good choice, because the distributions of profits or sales are naturally skewed. Simple graphical tools for the inspection of distributions are introduced in the next section.

### *1.7.1 Simple Graphics*

The degree of skewness of a distribution can be investigated by constructing histograms using the hist function. (More sophisticated alternatives such as smooth density estimates will be considered in Chapter 7.) For example, the code for producing Figure 1.1 first divides the plot region into two equally spaced rows (the layout function) and then plots the histograms of the raw market values in the upper part using the hist function. The lower part of

```
R> layout(matrix(1:2, nrow = 2))
R> hist(Forbes2000$marketvalue)
R> hist(log(Forbes2000$marketvalue))
```

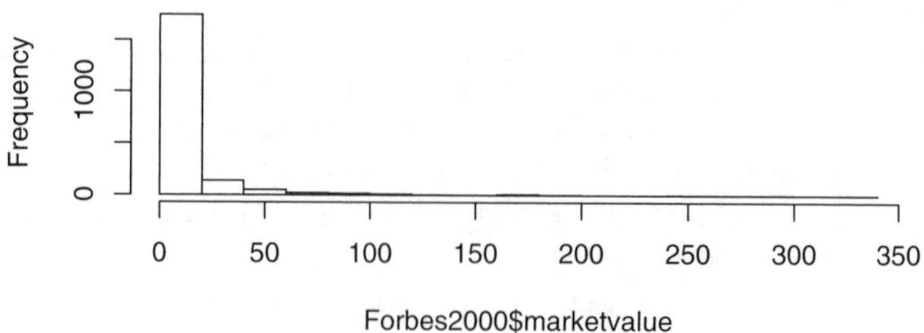


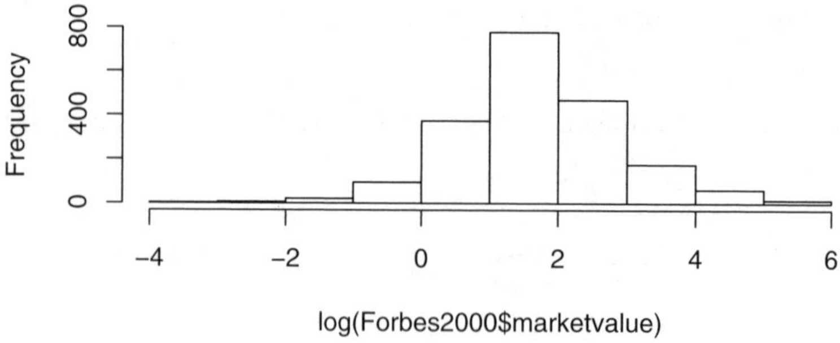


**Figure 1.1** Histograms of the market value and the logarithm of the market value for the companies contained in the Forbes 2000 list.

the figure depicts the histogram for the log transformed market values which appear to be more symmetric.

Bivariate relationships of two continuous variables are usually depicted as scatterplots. In R, regression relationships are specified by so-called *model formulae* which, in a simple bivariate case, may look like

```
R> fm <- marketvalue ~ sales
R> class(fm)

[1] "formula"
```

with the dependent variable on the left hand side and the independent variable on the right hand side. The tilde separates left and right hand side. Such

```
R> plot(log(marketvalue) ~ log(sales), data = Forbes2000,
+      pch = ".")
```

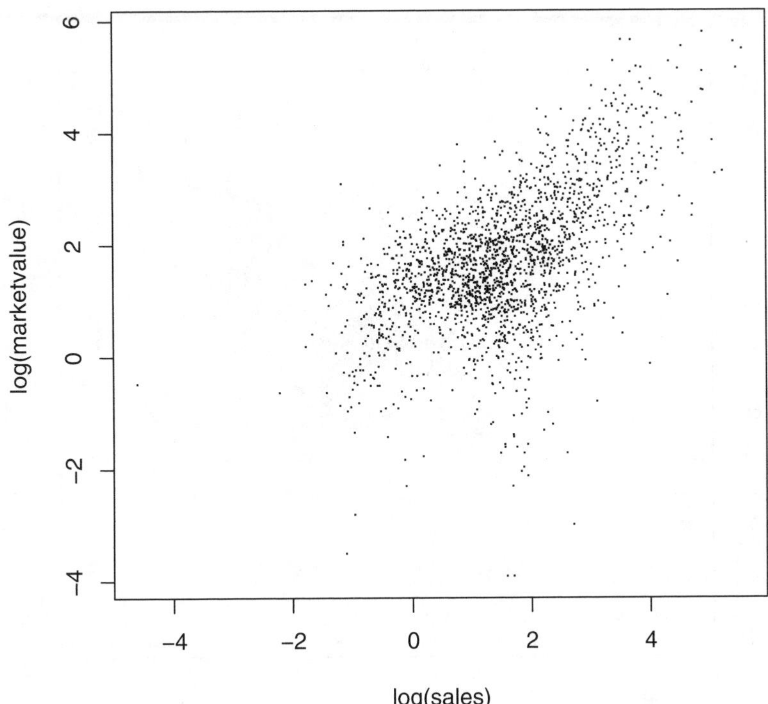


**Figure 1.2** Raw scatterplot of the logarithms of market value and sales.

a model formula can be passed to a model function (for example to the linear model function as explained in Chapter 5). The `plot` generic function implements a *formula* method as well. Because the distributions of both market value and sales are skewed we choose to depict their logarithms. A raw scatterplot of 2000 data points (Figure 1.2) is rather uninformative due to areas with very high density. This problem can be avoided by choosing a transparent color for the dots (currently only possible with the PDF graphics device) as shown in Figure 1.3.

If the independent variable is a factor, a boxplot representation is a natural choice. For four selected countries, the distributions of the logarithms of the market value may be visually compared in Figure 1.4. Here, the width of the boxes are proportional to the square root of the number of companies for each

```
R> pdf("figures/marketvalue-sales.pdf", version = "1.4")
R> plot(log(marketvalue) ~ log(sales), data = Forbes2000,
+      col = rgb(0, 0, 0, 0.1), pch = 16)
R> dev.off()
```

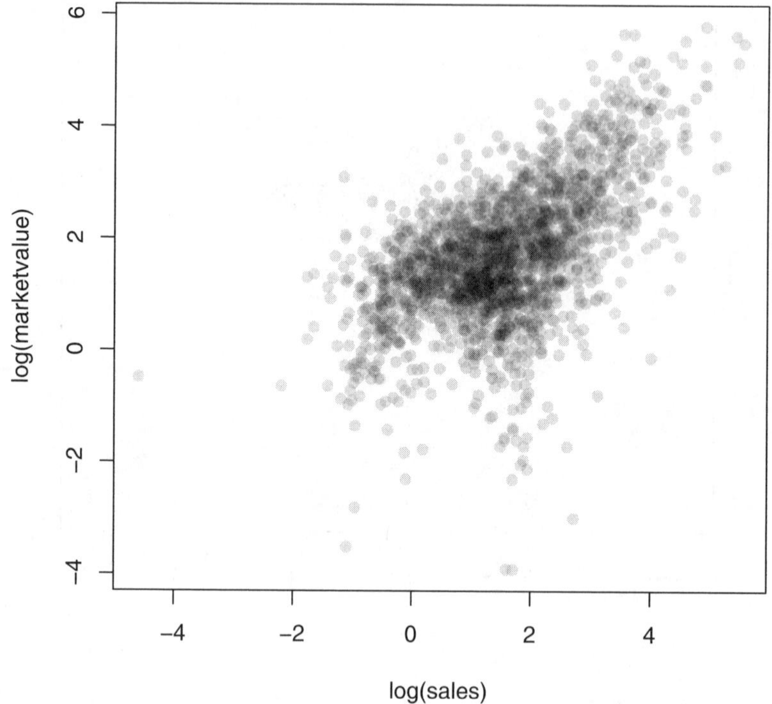


**Figure 1.3** Scatterplot with transparent shading of points of the logarithms of market value and sales.

country and extremely large or small market values are depicted by single points.

## 1.8 Organising an Analysis

Although it is possible to perform an analysis typing all commands directly on the R prompt it is much more comfortable to maintain a separate text file collecting all steps necessary to perform a certain data analysis task. Such an R transcript file, for example `analysis.R` created with your favourite text editor, can be sourced into R using the `source` command

```
R> boxplot(log(marketvalue) ~ country,
+      data = subset(Forbes2000,
+      country %in% c("United Kingdom", "Germany",
+          "India", "Turkey")), ylab = "log(marketvalue)",
+      varwidth = TRUE)
```

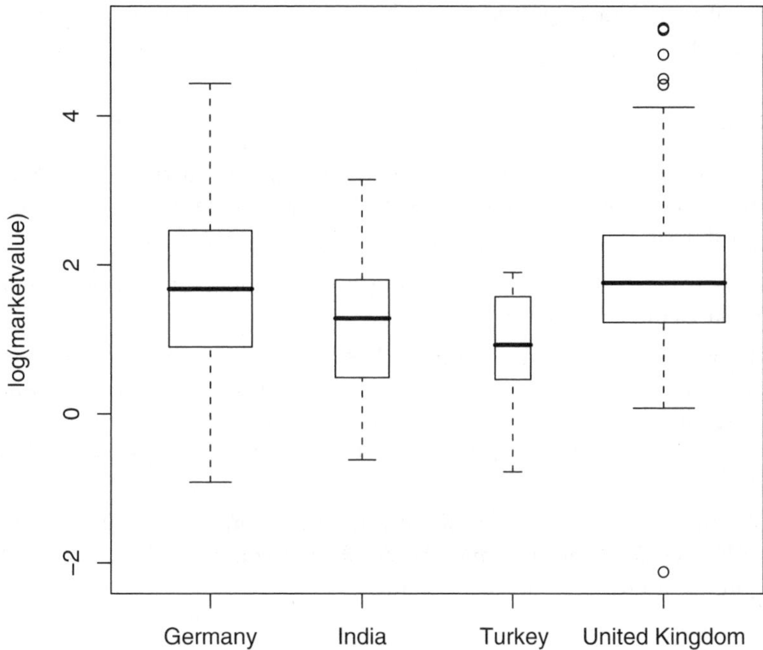


**Figure 1.4** Boxplots of the logarithms of the market value for four selected countries, the width of the boxes is proportional to the square-roots of the number of companies.

```
R> source("analysis.R", echo = TRUE)
```

When all steps of a data analysis, i.e., data preprocessing, transformations, simple summary statistics and plots, model building and inference as well as reporting, are collected in such an R transcript file, the analysis can be reproduced at any time, maybe with modified data as it frequently happens in our consulting practice.

## 1.9 Summary

Reading data into R is possible in many different ways, including direct connections to data base engines. Tabular data are handled by *data.frames* in R, and the usual data manipulation techniques such as sorting, ordering or subsetting can be performed by simple R statements. An overview on data stored in a *data.frame* is given mainly by two functions: `summary` and `str`. Simple graphics such as histograms and scatterplots can be constructed by applying the appropriate R functions (`hist` and `plot`) and we shall give many more examples of these functions and those that produce more interesting graphics in later chapters.

### Exercises

Ex. 1.1 Calculate the median profit for the companies in the United States and the median profit for the companies in the UK, France and Germany.

Ex. 1.2 Find all German companies with negative profit.

Ex. 1.3 Which business category are most of the companies situated at the Bermuda island working in?

Ex. 1.4 For the 50 companies in the Forbes data set with the highest profits, plot sales against assets (or some suitable transformation of each variable), labelling each point with the appropriate country name which may need to be abbreviated (using `abbreviate`) to avoid making the plot look too 'messy'.

Ex. 1.5 Find the average value of sales for the companies in each country in the Forbes data set, and find the number of companies in each country with profits above 5 billion US dollars.

CHAPTER 2

# Simple Inference: Guessing Lengths, Wave Energy, Water Hardness, Piston Rings, and Rearrests of Juveniles

## 2.1 Introduction

Shortly after metric units of length were officially introduced in Australia in the 1970s, each of a group of 44 students was asked to guess, to the nearest metre, the width of the lecture hall in which they were sitting. Another group of 69 students in the same room was asked to guess the width in feet, to the nearest foot. The data were collected by Professor T. Lewis, and are given here in Table 2.1, which is taken from Hand et al. (1994). The main question is whether estimation in feet and in metres gives different results.

**Table 2.1**: `roomwidth` data. Room width estimates (`width`) in feet and in metres (`unit`).

| unit | width | unit | width | unit | width | unit | width |
|---|---|---|---|---|---|---|---|
| metres | 8 | metres | 16 | feet | 34 | feet | 45 |
| metres | 9 | metres | 16 | feet | 35 | feet | 45 |
| metres | 10 | metres | 17 | feet | 35 | feet | 45 |
| metres | 10 | metres | 17 | feet | 36 | feet | 45 |
| metres | 10 | metres | 17 | feet | 36 | feet | 45 |
| metres | 10 | metres | 17 | feet | 36 | feet | 46 |
| metres | 10 | metres | 18 | feet | 37 | feet | 46 |
| metres | 10 | metres | 18 | feet | 37 | feet | 47 |
| metres | 11 | metres | 20 | feet | 40 | feet | 48 |
| metres | 11 | metres | 22 | feet | 40 | feet | 48 |
| metres | 11 | metres | 25 | feet | 40 | feet | 50 |
| metres | 11 | metres | 27 | feet | 40 | feet | 50 |
| metres | 12 | metres | 35 | feet | 40 | feet | 50 |
| metres | 12 | metres | 38 | feet | 40 | feet | 51 |
| metres | 13 | metres | 40 | feet | 40 | feet | 54 |
| metres | 13 | feet | 24 | feet | 40 | feet | 54 |
| metres | 13 | feet | 25 | feet | 40 | feet | 54 |
| metres | 14 | feet | 27 | feet | 41 | feet | 55 |
| metres | 14 | feet | 30 | feet | 41 | feet | 55 |
| metres | 14 | feet | 30 | feet | 42 | feet | 60 |

**Table 2.1**: `roomwidth` data (continued).

| unit | width | unit | width | unit | width | unit | width |
|---|---|---|---|---|---|---|---|
| metres | 15 | feet | 30 | feet | 42 | feet | 60 |
| metres | 15 | feet | 30 | feet | 42 | feet | 63 |
| metres | 15 | feet | 30 | feet | 42 | feet | 70 |
| metres | 15 | feet | 30 | feet | 43 | feet | 75 |
| metres | 15 | feet | 32 | feet | 43 | feet | 80 |
| metres | 15 | feet | 32 | feet | 44 | feet | 94 |
| metres | 15 | feet | 33 | feet | 44 | | |
| metres | 15 | feet | 34 | feet | 44 | | |
| metres | 16 | feet | 34 | feet | 45 | | |

In a design study for a device to generate electricity from wave power at sea, experiments were carried out on scale models in a wave tank to establish how the choice of mooring method for the system affected the bending stress produced in part of the device. The wave tank could simulate a wide range of sea states and the model system was subjected to the same sample of sea states with each of two mooring methods, one of which was considerably cheaper than the other. The resulting data (from Hand et al., 1994, giving root mean square bending moment in Newton metres) are shown in Table 2.2. The question of interest is whether bending stress differs for the two mooring methods.

**Table 2.2**: `waves` data. Bending stress (root mean squared bending moment in Newton metres) for two mooring methods in a wave energy experiment.

| method1 | method2 | method1 | method2 | method1 | method2 |
|---|---|---|---|---|---|
| 2.23 | 1.82 | 8.98 | 8.88 | 5.91 | 6.44 |
| 2.55 | 2.42 | 0.82 | 0.87 | 5.79 | 5.87 |
| 7.99 | 8.26 | 10.83 | 11.20 | 5.50 | 5.30 |
| 4.09 | 3.46 | 1.54 | 1.33 | 9.96 | 9.82 |
| 9.62 | 9.77 | 10.75 | 10.32 | 1.92 | 1.69 |
| 1.59 | 1.40 | 5.79 | 5.87 | 7.38 | 7.41 |

The data shown in Table 2.3 were collected in an investigation of environmental causes of disease and are taken from Hand et al. (1994). They show the annual mortality per 100,000 for males, averaged over the years 1958–1964, and the calcium concentration (in parts per million) in the drinking water for 61 large towns in England and Wales. The higher the calcium concentration, the harder the water. Towns at least as far north as Derby are identified in the table. Here there are several questions that might be of interest including: are mortality and water hardness related, and do either or both variables differ between northern and southern towns?

**Table 2.3**: `water` data. Mortality (per 100,000 males per year, `mortality`) and water hardness for 61 cities in England and Wales.

| `location` | `town` | `mortality` | `hardness` |
|---|---:|---:|---:|
| South | Bath | 1247 | 105 |
| North | Birkenhead | 1668 | 17 |
| South | Birmingham | 1466 | 5 |
| North | Blackburn | 1800 | 14 |
| North | Blackpool | 1609 | 18 |
| North | Bolton | 1558 | 10 |
| North | Bootle | 1807 | 15 |
| South | Bournemouth | 1299 | 78 |
| North | Bradford | 1637 | 10 |
| South | Brighton | 1359 | 84 |
| South | Bristol | 1392 | 73 |
| North | Burnley | 1755 | 12 |
| South | Cardiff | 1519 | 21 |
| South | Coventry | 1307 | 78 |
| South | Croydon | 1254 | 96 |
| North | Darlington | 1491 | 20 |
| North | Derby | 1555 | 39 |
| North | Doncaster | 1428 | 39 |
| South | East Ham | 1318 | 122 |
| South | Exeter | 1260 | 21 |
| North | Gateshead | 1723 | 44 |
| North | Grimsby | 1379 | 94 |
| North | Halifax | 1742 | 8 |
| North | Huddersfield | 1574 | 9 |
| North | Hull | 1569 | 91 |
| South | Ipswich | 1096 | 138 |
| North | Leeds | 1591 | 16 |
| South | Leicester | 1402 | 37 |
| North | Liverpool | 1772 | 15 |
| North | Manchester | 1828 | 8 |
| North | Middlesbrough | 1704 | 26 |
| North | Newcastle | 1702 | 44 |
| South | Newport | 1581 | 14 |
| South | Northampton | 1309 | 59 |
| South | Norwich | 1259 | 133 |
| North | Nottingham | 1427 | 27 |
| North | Oldham | 1724 | 6 |

**Table 2.3**: `water` data (continued).

| `location` | `town` | `mortality` | `hardness` |
|---|---:|---:|---:|
| South | Oxford | 1175 | 107 |
| South | Plymouth | 1486 | 5 |
| South | Portsmouth | 1456 | 90 |
| North | Preston | 1696 | 6 |
| South | Reading | 1236 | 101 |
| North | Rochdale | 1711 | 13 |
| North | Rotherham | 1444 | 14 |
| North | St Helens | 1591 | 49 |
| North | Salford | 1987 | 8 |
| North | Sheffield | 1495 | 14 |
| South | Southampton | 1369 | 68 |
| South | Southend | 1257 | 50 |
| North | Southport | 1587 | 75 |
| North | South Shields | 1713 | 71 |
| North | Stockport | 1557 | 13 |
| North | Stoke | 1640 | 57 |
| North | Sunderland | 1709 | 71 |
| South | Swansea | 1625 | 13 |
| North | Wallasey | 1625 | 20 |
| South | Walsall | 1527 | 60 |
| South | West Bromwich | 1627 | 53 |
| South | West Ham | 1486 | 122 |
| South | Wolverhampton | 1485 | 81 |
| North | York | 1378 | 71 |

The two-way contingency table in Table 2.4 shows the number of piston-ring failures in each of three legs of four steam-driven compressors located in the same building (Haberman, 1973). The compressors have identical design and are oriented in the same way. The question of interest is whether the two categorical variables (compressor and leg) are independent.

The data in Table 2.5 (taken from Agresti, 1996) arise from a sample of juveniles convicted of felony in Florida in 1987. Matched pairs were formed using criteria such as age and the number of previous offences. For each pair, one subject was handled in the juvenile court and the other was transferred to the adult court. Whether or not the juvenile was rearrested by the end of 1988 was then noted. Here the question of interest is whether the true proportions rearrested were identical for the adult and juvenile court assignments?

**Table 2.4**: `pistonrings` data. Number of piston ring failures for three legs of four compressors.

| compressor | | leg | | |
|---|---|---|---|---|
| | | North | Centre | South |
| | C1 | 17 | 17 | 12 |
| | C2 | 11 | 9 | 13 |
| | C3 | 11 | 8 | 19 |
| | C4 | 14 | 7 | 28 |

*Source*: From Haberman, S. J., *Biometrics*, 29, 205–220, 1973. With permission.

**Table 2.5**: `rearrests` data. Rearrests of juvenile felons by type of court in which they were tried.

| Adult court | | Juvenile court | |
|---|---|---|---|
| | | Rearrest | No rearrest |
| | Rearrest | 158 | 515 |
| | No rearrest | 290 | 1134 |

*Source*: From Agresti, A., *An Introduction to Categorical Data Analysis*, John Wiley & Sons, New York, 1996. With permission.

## 2.2 Statistical Tests

Inference, the process of drawing conclusions about a population on the basis of measurements or observations made on a sample of individuals from the population, is central to statistics. In this chapter we shall use the data sets described in the introduction to illustrate both the application of the most common statistical tests, and some simple graphics that may often be used to aid in understanding the results of the tests. Brief descriptions of each of the tests to be used follow.

### *2.2.1 Comparing Normal Populations: Student's t-Tests*

The $t$-test is used to assess hypotheses about two population means where the measurements are assumed to be sampled from a normal distribution. We shall describe two types of $t$-tests, the independent samples test and the paired test.

The independent samples $t$-test is used to test the null hypothesis that the means of two populations are the same, $H_0 : \mu_1 = \mu_2$, when a sample of observations from each population is available. The subjects of one population must not be individually matched with subjects from the other population and the subjects within each group should not be related to each other. The variable to be compared is assumed to have a normal distribution with the same standard deviation in both populations. The test statistic is essentially a standardised difference of the two sample means,

$$t = \frac{\bar{y}_1 - \bar{y}_2}{s\sqrt{1/n_1 + 1/n_2}} \tag{2.1}$$

where $\bar{y}_i$ and $n_i$ are the means and sample sizes in groups $i = 1$ and 2, respectively. The pooled standard deviation $s$ is given by

$$s = \sqrt{\frac{(n_1 - 1)s_1^2 + (n_2 - 1)s_2^2}{n_1 + n_2 - 2}}$$

where $s_1$ and $s_2$ are the standard deviations in the two groups.

Under the null hypothesis, the $t$-statistic has a Student's $t$-distribution with $n_1 + n_2 - 2$ degrees of freedom. A $100(1 - \alpha)\%$ confidence interval for the difference between two means is useful in giving a plausible range of values for the differences in the two means and is constructed as

$$\bar{y}_1 - \bar{y}_2 \pm t_{\alpha, n_1 + n_2 - 2}\, s\sqrt{n_1^{-1} + n_2^{-1}}$$

where $t_{\alpha, n_1 + n_2 - 2}$ is the percentage point of the $t$-distribution such that the cumulative distribution function, $\mathsf{P}(t \leq t_{\alpha, n_1 + n_2 - 2})$, equals $1 - \alpha/2$.

If the two populations are suspected of having different variances, a modified form of the $t$ statistic, known as the Welch test, may be used, namely

$$t = \frac{\bar{y}_1 - \bar{y}_2}{\sqrt{s_1^2/n_1 + s_2^2/n_2}}.$$

In this case, $t$ has a Student's $t$-distribution with $\nu$ degrees of freedom, where

$$\nu = \left( \frac{c}{n_1 - 1} + \frac{(1 - c)^2}{n_2 - 1} \right)^{-1}$$

with

$$c = \frac{s_1^2/n_1}{s_1^2/n_1 + s_2^2/n_2}.$$

A paired $t$-test is used to compare the means of two populations when samples from the populations are available, in which each individual in one sample is paired with an individual in the other sample or each individual in the sample is observed twice. Examples of the former are anorexic girls and their healthy sisters and of the latter the same patients observed before and after treatment.

If the values of the variable of interest, $y$, for the members of the $i$th pair in groups 1 and 2 are denoted as $y_{1i}$ and $y_{2i}$, then the differences $d_i = y_{1i} - y_{2i}$ are assumed to have a normal distribution with mean $\mu$ and the null hypothesis here is that the mean difference is zero, i.e., $H_0 : \mu = 0$. The paired $t$-statistic is

$$t = \frac{\bar{d}}{s/\sqrt{n}}$$

where $\bar{d}$ is the mean difference between the paired measurements and $s$ is its standard deviation. Under the null hypothesis, $t$ follows a $t$-distribution with $n - 1$ degrees of freedom. A $100(1 - \alpha)\%$ confidence interval for $\mu$ can be constructed by

$$\bar{d} \pm t_{\alpha,n-1} s/\sqrt{n}$$

where $\mathsf{P}(t \leq t_{\alpha,n-1}) = 1 - \alpha/2$.

### *2.2.2 Non-parametric Analogues of Independent Samples and Paired t-Tests*

One of the assumptions of both forms of $t$-test described above is that the data have a normal distribution, i.e., are unimodal and symmetric. When departures from those assumptions are extreme enough to give cause for concern, then it might be advisable to use the non-parametric analogues of the $t$-tests, namely the *Wilcoxon Mann-Whitney rank sum test* and the *Wilcoxon signed rank test.* In essence, both procedures throw away the original measurements and only retain the rankings of the observations.

For two independent groups, the Wilcoxon Mann-Whitney rank sum test applies the $t$-statistic to the joint ranks of all measurements in both groups instead of the original measurements. The null hypothesis to be tested is that the two populations being compared have identical distributions. For two normally distributed populations with common variance, this would be equivalent to the hypothesis that the means of the two populations are the same. The alternative hypothesis is that the population distributions differ in location, i.e., the median.

The test is based on the joint ranking of the observations from the two samples (as if they were from a single sample). The test statistic is the sum of the ranks of one sample (the lower of the two rank sums is generally used). A version of this test applicable in the presence of ties is discussed in Chapter 3.

For small samples, $p$-values for the test statistic can be assigned relatively simply. A large sample approximation is available that is suitable when the two sample sizes are greater and there are no ties. In R, the large sample approximation is used by default when the sample size in one group exceeds 50 observations.

In the paired situation, we first calculate the differences $d_i = y_{1i} - y_{2i}$ between each pair of observations. To compute the Wilcoxon signed-rank statistic, we rank the absolute differences $|d_i|$. The statistic is defined as the sum of the ranks associated with positive difference $d_i > 0$. Zero differences are discarded, and the sample size $n$ is altered accordingly. Again, $p$-values for small sample sizes can be computed relatively simply and a large sample approximation is available. It should be noted that this test is only valid when the differences $d_i$ are symmetrically distributed.

### *2.2.3 Testing Independence in Contingency Tables*

When a sample of $n$ observations in two nominal (categorical) variables are available, they can be arranged into a cross-classification (see Table 2.6) in which the number of observations falling in each cell of the table is recorded. Table 2.6 is an example of such a contingency table, in which the observations for a sample of individuals or objects are cross-classified with respect to two categorical variables. Testing for the independence of the two variables $x$ and $y$ is of most interest in general and details of the appropriate test follow.

**Table 2.6**: The general $r \times c$ table.

| | | $y$ | | | |
|---|---|---|---|---|---|
| | | 1 | ... | $c$ | |
| $x$ | 1 | $n_{11}$ | ... | $n_{1c}$ | $n_{1\cdot}$ |
| | 2 | $n_{21}$ | ... | $n_{2c}$ | $n_{2\cdot}$ |
| | ⋮ | ⋮ | ... | ⋮ | ⋮ |
| | $r$ | $n_{r1}$ | ... | $n_{rc}$ | $n_{r\cdot}$ |
| | | $n_{\cdot 1}$ | ... | $n_{\cdot c}$ | $n$ |

Under the null hypothesis of independence of the row variable $x$ and the column variable $y$, estimated expected values $E_{jk}$ for cell $(j, k)$ can be computed from the corresponding margin totals $E_{jk} = n_{j\cdot}n_{\cdot k}/n$. The test statistic for assessing independence is

$$X^2 = \sum_{j=1}^{r} \sum_{k=1}^{c} \frac{(n_{jk} - E_{jk})^2}{E_{jk}}.$$

Under the null hypothesis of independence, the test statistic $X^2$ is asymptotically distributed according to a $\chi^2$-distribution with $(r-1)(c-1)$ degrees of freedom, the corresponding test is usually known as *chi-squared test.*

### *2.2.4 McNemar's Test*

The chi-squared test on categorical data described previously assumes that the observations are independent. Often, however, categorical data arise from *paired* observations, for example, cases matched with controls on variables such as sex, age and so on, or observations made on the same subjects on two occasions (cf. paired $t$-test). For this type of paired data, the required procedure is McNemar's test. The general form of such data is shown in Table 2.7.

**Table 2.7**: Frequencies in matched samples data.

| | | Sample 1 | |
|---|---|---|---|
| | | present | absent |
| Sample 2 | present | $a$ | $b$ |
| | absent | $c$ | $d$ |

Under the hypothesis that the two populations do not differ in their probability of having the characteristic present, the test statistic

$$X^2 = \frac{(c-b)^2}{c+b}$$

has a $\chi^2$-distribution with a single degree of freedom.

## 2.3 Analysis Using R

### 2.3.1 Estimating the Width of a Room

The data shown in Table 2.1 are available as `roomwidth` *data.frame* from the *HSAUR* package and can be attached by using

```
R> data("roomwidth", package = "HSAUR")
```

If we convert the estimates of the room width in metres into feet by multiplying each by 3.28 then we would like to test the hypothesis that the mean of the population of 'metre' estimates is equal to the mean of the population of 'feet' estimates. We shall do this first by using an independent samples $t$-test, but first it is good practice to, informally at least, check the normality and equal variance assumptions. Here we can use a combination of numerical and graphical approaches. The first step should be to convert the metre estimates into feet, i.e., by a factor

```
R> convert <- ifelse(roomwidth$unit == "feet", 1, 3.28)
```

which equals one for all feet measurements and 3.28 for the measurements in metres. Now, we get the usual summary statistics and standard deviations of each set of estimates using

```
R> tapply(roomwidth$width * convert, roomwidth$unit,
+      summary)

$feet
   Min. 1st Qu.  Median    Mean 3rd Qu.    Max. 
   24.0    36.0    42.0    43.7    48.0    94.0 

$metres
```

```
   Min. 1st Qu.  Median    Mean 3rd Qu.    Max.
  26.24   36.08   49.20   52.55   55.76  131.20
R> tapply(roomwidth$width * convert, roomwidth$unit,
+      sd)
    feet   metres
12.49742 23.43444
```

where tapply applies summary, or sd, to the converted widths for both groups of measurements given by roomwidth$unit. A boxplot of each set of estimates might be useful and is depicted in Figure 2.1. The layout function (line 1 in Figure 2.1) divides the plotting area in three parts. The boxplot function produces a boxplot in the upper part and the two qqnorm statements in lines 8 and 11 set up the normal probability plots that can be used to assess the normality assumption of the $t$-test.

The boxplots indicate that both sets of estimates contain a number of outliers and also that the estimates made in metres are skewed and more variable than those made in feet, a point underlined by the numerical summary statistics above. Both normal probability plots depart from linearity, suggesting that the distributions of both sets of estimates are not normal. The presence of outliers, the apparently different variances and the evidence of non-normality all suggest caution in applying the $t$-test, but for the moment we shall apply the usual version of the test using the t.test function in R.

The two-sample test problem is specified by a *formula*, here by

```
I(width * convert) ~ unit
```

where the response, width, on the left hand side needs to be converted first and, because the star has a special meaning in formulae as will be explained in Chapter 4, the conversion needs to be embedded by I. The factor unit on the right hand side specifies the two groups to be compared.

From the output shown in Figure 2.2 we see that there is considerable evidence that the estimates made in feet are lower than those made in metres by between about 2 and 15 feet. The test statistic $t$ from 2.1 is $-2.615$ and, with 111 degrees of freedom, the two-sided $p$-value is 0.01. In addition, a 95% confidence interval for the difference of the estimated widths between feet and metres is reported.

But this form of $t$-test assumes both normality and equality of population variances, both of which are suspect for these data. Departure from the equality of variance assumption can be accommodated by the modified $t$-test described above and this can be applied in R by choosing var.equal = FALSE (note that var.equal = FALSE is the default in R). The result shown in Figure 2.3 as well indicates that there is strong evidence for a difference in the means of the two types of estimate.

But there remains the problem of the outliers and the possible non-normality; consequently we shall apply the Wilcoxon Mann-Whitney test which since it is based on the ranks of the observations is unlikely to be affected by the outliers, and which does not assume that the data have a normal distribution. The test can be applied in R using the wilcox.test function.

```
R> layout(matrix(c(1, 2, 1, 3), nrow = 2, ncol = 2,
+      byrow = FALSE))
R> boxplot(I(width * convert) ~ unit, data = roomwidth,
+      ylab = "Estimated width (feet)", varwidth = TRUE,
+      names = c("Estimates in feet",
+      "Estimates in metres (converted to feet)"))
R> feet <- roomwidth$unit == "feet"
R> qqnorm(roomwidth$width[feet],
+      ylab = "Estimated width (feet)")
R> qqline(roomwidth$width[feet])
R> qqnorm(roomwidth$width[!feet],
+      ylab = "Estimated width (metres)")
R> qqline(roomwidth$width[!feet])
```

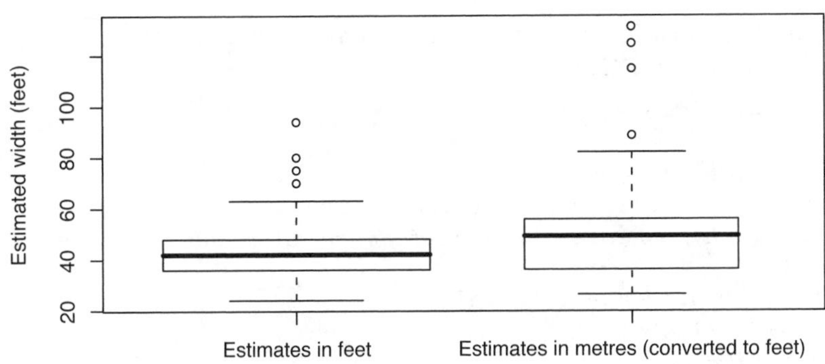


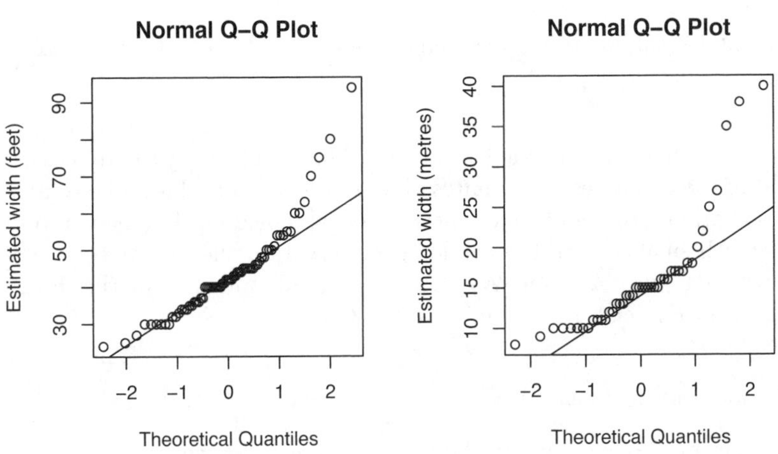


**Figure 2.1** Boxplots of estimates of width of room in feet and metres (after conversion to feet) and normal probability plots of estimates of room width made in feet and in metres.

```
R> t.test(I(width * convert) ~ unit, data = roomwidth,
+      var.equal = TRUE)

        Two Sample t-test

data:  I(width * convert) by unit
t = -2.6147, df = 111, p-value = 0.01017
alternative hypothesis: true difference in means is not equal
    to 0
95 percent confidence interval:
 -15.572734  -2.145052
sample estimates:
  mean in group feet mean in group metres
            43.69565             52.55455
```

**Figure 2.2** R output of the independent samples $t$-test for the `roomwidth` data.

```
R> t.test(I(width * convert) ~ unit, data = roomwidth,
+      var.equal = FALSE)

        Welch Two Sample t-test

data:  I(width * convert) by unit
t = -2.3071, df = 58.788, p-value = 0.02459
alternative hypothesis: true difference in means is not equal
    to 0
95 percent confidence interval:
 -16.54308  -1.17471
sample estimates:
  mean in group feet mean in group metres
            43.69565             52.55455
```

**Figure 2.3** R output of the independent samples Welch test for the `roomwidth` data.

Figure 2.4 shows a two-sided $p$-value of 0.028 confirming the difference in location of the two types of estimates of room width. Note that, due to ranking the observations, the confidence interval for the median difference reported here is much smaller than the confidence interval for the difference in means as shown in Figures 2.2 and 2.3. Further possible analyses of the data are considered in Exercise 2.1 and in Chapter 3.

### *2.3.2 Wave Energy Device Mooring*

The data from Table 2.2 are available as *data.frame* `waves`

```
R> data("waves", package = "HSAUR")
```

```
R> wilcox.test(I(width * convert) ~ unit, data = roomwidth,
+     conf.int = TRUE)

        Wilcoxon rank sum test with continuity correction

data:  I(width * convert) by unit
W = 1145, p-value = 0.02815
alternative hypothesis: true mu is not equal to 0
95 percent confidence interval:
 -9.3599953 -0.8000423
sample estimates:
difference in location
             -5.279955
```

**Figure 2.4** R output of the Wilcoxon rank sum test for the `roomwidth` data.

and requires the use of a matched pairs $t$-test to answer the question of interest. This test assumes that the differences between the matched observations have a normal distribution so we can begin by checking this assumption by constructing a boxplot and a normal probability plot – see Figure 2.5.

The boxplot indicates a possible outlier, and the normal probability plot gives little cause for concern about departures from normality, although with only 18 observations it is perhaps difficult to draw any convincing conclusion. We can now apply the paired $t$-test to the data again using the `t.test` function. Figure 2.6 shows that there is no evidence for a difference in the mean bending stress of the two types of mooring device. Although there is no real reason for applying the non-parametric analogue of the paired $t$-test to these data, we give the R code for interest in Figure 2.7. The associated $p$-value is 0.316 confirming the result from the $t$-test.

### *2.3.3 Mortality and Water Hardness*

There is a wide range of analyses we could apply to the data in Table 2.3 available from

```
R> data("water", package = "HSAUR")
```

But to begin we will construct a scatterplot of the data enhanced somewhat by the addition of information about the marginal distributions of water hardness (calcium concentration) and mortality, and by adding the estimated linear regression fit (see Chapter 5) for mortality on hardness. The plot and the required R code is given along with Figure 2.8. In line 1 of Figure 2.8, we divide the plotting region into four areas of different size. The scatterplot (line 3) uses a plotting symbol depending on the location of the city (by the `pch` argument), a legend for the location is added in line 6. We add a least squares fit (see Chapter 5) to the scatterplot and, finally, depict the marginal distributions by means of a boxplot and a histogram. The scatterplot shows

```
R> mooringdiff <- waves$method1 - waves$method2
R> layout(matrix(1:2, ncol = 2))
R> boxplot(mooringdiff, ylab = "Differences (Newton metres)",
+      main = "Boxplot")
R> abline(h = 0, lty = 2)
R> qqnorm(mooringdiff, ylab = "Differences (Newton metres)")
R> qqline(mooringdiff)
```

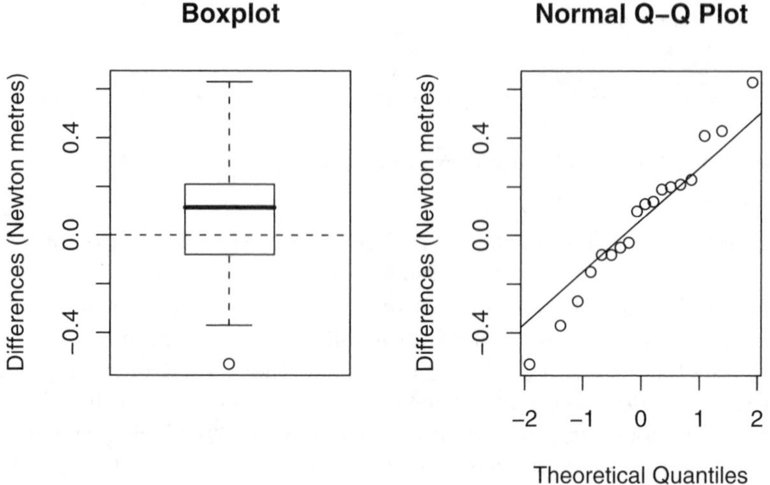


**Figure 2.5** Boxplot and normal probability plot for differences between the two mooring methods.

that as hardness increases mortality decreases, and the histogram for the water hardness shows it has a rather skewed distribution.

We can both calculate the Pearson's correlation coefficient between the two variables and test whether it differs significantly for zero by using the `cor.test` function in R. The test statistic for assessing the hypothesis that the population correlation coefficient is zero is

$$r/\sqrt{(1-r^2)/(n-2)}$$

where $r$ is the sample correlation coefficient and $n$ is the sample size. If the population correlation is zero and assuming the data have a bivariate normal distribution, then the test statistic has a Student's $t$ distribution with $n-2$ degrees of freedom.

The estimated correlation shown in Figure 2.9 is -0.655 and is highly significant. We might also be interested in the correlation between water hardness

```
R> t.test(mooringdiff)

        One Sample t-test

data:  mooringdiff
t = 0.9019, df = 17, p-value = 0.3797
alternative hypothesis: true mean is not equal to 0
95 percent confidence interval:
 -0.08258476  0.20591810
sample estimates:
 mean of x
0.06166667
```

**Figure 2.6** R output of the paired $t$-test for the `waves` data.

```
R> wilcox.test(mooringdiff)

        Wilcoxon signed rank test with continuity correction

data:  mooringdiff
V = 109, p-value = 0.3165
alternative hypothesis: true mu is not equal to 0
```

**Figure 2.7** R output of the Wilcoxon signed rank test for the `waves` data.

and mortality in each of the regions North and South but we leave this as an exercise for the reader (see Exercise 2.2).

### *2.3.4 Piston-ring Failures*

The first step in the analysis of the `pistonrings` data is to apply the chi-squared test for independence. This we can do in R using the `chisq.test` function. The output of the chi-squared test, see Figure 2.10, shows a value of the $X^2$ test statistic of 11.722 with 6 degrees of freedom and an associated $p$-value of 0.068. The evidence for departure from independence of compressor and leg is not strong, but it may be worthwhile taking the analysis a little further by examining the estimated expected values and the differences of these from the corresponding observed value.

Rather than looking at the simple differences of observed and expected values for each cell which would be unsatisfactory since a difference of fixed size is clearly more important for smaller samples, it is preferable to consider a *standardised residual* given by dividing the observed minus expected difference by the square root of the appropriate expected value. The $X^2$ statistic for assessing independence is simply the sum, over all the cells in the table, of the squares of these terms. We can find these values extracting the `residuals` element of the object returned by the `chisq.test` function

```
R> nf <- layout(matrix(c(2, 0, 1, 3), 2, 2, byrow = TRUE),
+     c(2, 1), c(1, 2), TRUE)
R> psymb <- as.numeric(water$location)
R> plot(mortality ~ hardness, data = water, pch = psymb)
R> abline(lm(mortality ~ hardness, data = water))
R> legend("topright", legend = levels(water$location),
+     pch = c(1, 2), bty = "n")
R> hist(water$hardness)
R> boxplot(water$mortality)
```

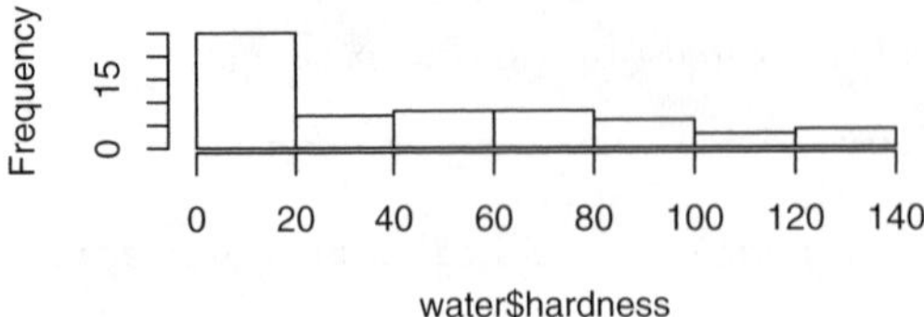


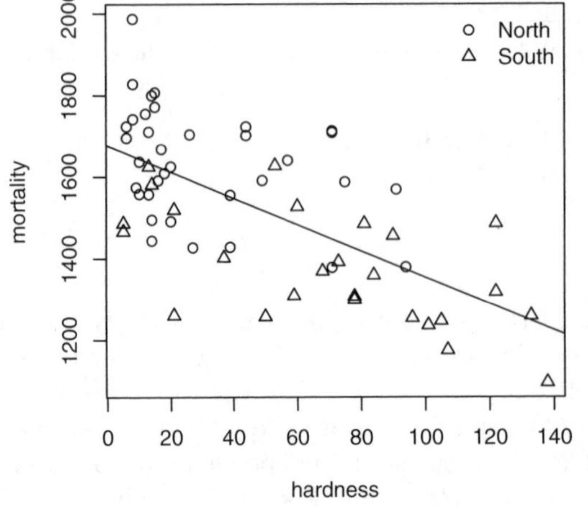


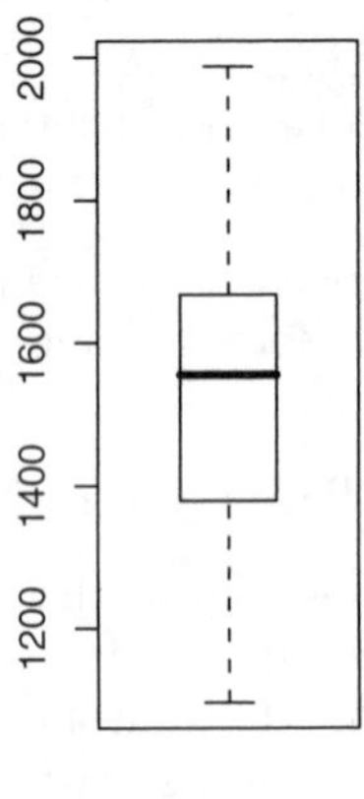


**Figure 2.8** Enhanced scatterplot of water hardness and mortality, showing both the joint and the marginal distributions and, in addition, the location of the city by different plotting symbols.

```
R> cor.test(~mortality + hardness, data = water)

        Pearson's product-moment correlation

data:  mortality and hardness
t = -6.6555, df = 59, p-value = 1.033e-08
alternative hypothesis: true correlation is not equal to 0
95 percent confidence interval:
 -0.7783208 -0.4826129
sample estimates:
       cor
-0.6548486
```

**Figure 2.9** R output of Pearsons' correlation coefficient for the `water` data.

```
R> data("pistonrings", package = "HSAUR")
R> chisq.test(pistonrings)

        Pearson's Chi-squared test

data:  pistonrings
X-squared = 11.7223, df = 6, p-value = 0.06846
```

**Figure 2.10** R output of the chi-squared test for the `pistonrings` data.

```
R> chisq.test(pistonrings)$residuals

          leg
compressor      North     Centre      South
        C1  0.6036154  1.6728267 -1.7802243
        C2  0.1429031  0.2975200 -0.3471197
        C3 -0.3251427 -0.4522620  0.6202463
        C4 -0.4157886 -1.4666936  1.4635235
```

A graphical representation of these residuals is called *association plot* and is available via the `assoc` function from package *vcd* (Meyer et al., 2005) applied to the contingency table of the two categorical variables. Figure 2.11 depicts the residuals for the piston ring data. The deviations from independence are largest for C1 and C4 compressors in the centre and south leg.

It is tempting to think that the size of these residuals may be judged by comparison with standard normal percentage points (for example greater than 1.96 or less than 1.96 for significance level $\alpha = 0.05$). Unfortunately it can be shown that the variance of a standardised residual is always less than or equal to one, and in some cases considerably less than one, however, the residuals are asymptotically normal. A more satisfactory 'residual' for contingency table data is considered in Exercise 2.3.

```
R> library("vcd")
R> assoc(pistonrings)
```

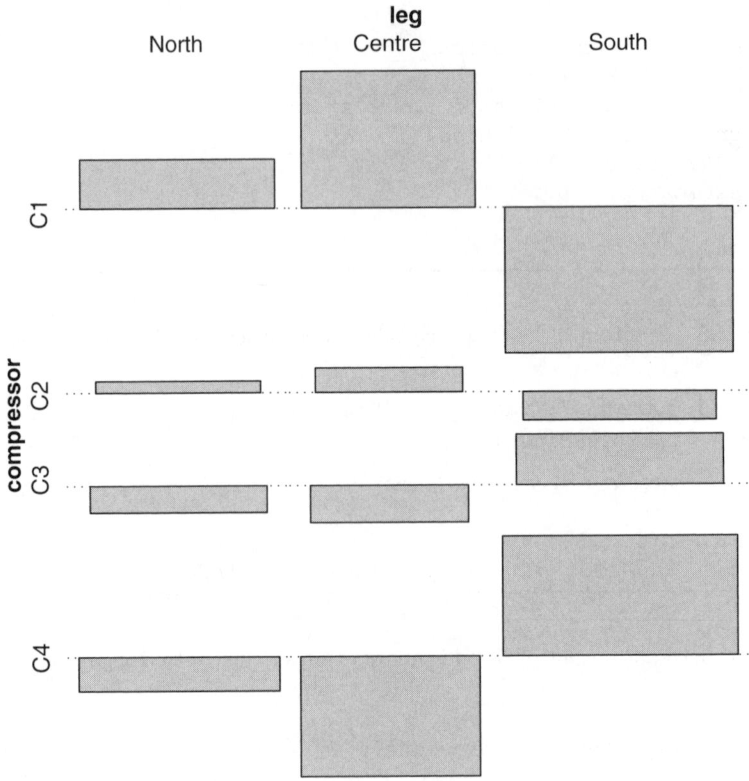


**Figure 2.11** Association plot of the residuals for the `pistonrings` data.

### *2.3.5 Rearrests of Juveniles*

The data in Table 2.5 are available as *table* object via

```
R> data("rearrests", package = "HSAUR")
R> rearrests

               Juvenile court
Adult court     Rearrest No rearrest
  Rearrest           158         515
  No rearrest        290        1134
```

and in `rearrests` the counts in the four cells refer to the matched pairs of subjects; for example, in 158 pairs both members of the pair were rearrested. Here we need to use McNemar's test to assess whether rearrest is associated with type of court where the juvenile was tried. We can use the R function `mcnemar.test`. The test statistic shown in Figure 2.12 is 62.888 with a single degree of freedom – the associated $p$-value is extremely small and there is strong evidence that type of court and the probability of rearrest are related. It appears that trial at a juvenile court is less likely to result in rearrest (see Exercise 2.4). An exact version of McNemar's test can be obtained by testing whether $b$ and $c$ are equal using a binomial test (see Figure 2.13).

```
R> mcnemar.test(rearrests, correct = FALSE)

        McNemar's Chi-squared test

data:  rearrests
McNemar's chi-squared = 62.8882, df = 1, p-value =
2.188e-15
```

**Figure 2.12** R output of McNemar's test for the `rearrests` data.

```
R> binom.test(rearrests[2], n = sum(rearrests[c(2,
+     3)]))

        Exact binomial test

data:  rearrests[2] and sum(rearrests[c(2, 3)])
number of successes = 290, number of trials = 805,
p-value = 1.918e-15
alternative hypothesis: true probability of success is not
    equal to 0.5
95 percent confidence interval:
 0.3270278 0.3944969
sample estimates:
probability of success
             0.3602484
```

**Figure 2.13** R output of an exact version of McNemar's test for the `rearrests` data computed via a binomal test.

## 2.4 Summary

Significance tests are widely used and they can easily be applied using the corresponding functions in R. But they often need to be accompanied by some graphical material to aid in interpretation and to assess whether assumptions are met. In addition, $p$-values are never as useful as confidence intervals.

**Exercises**

Ex. 2.1 After the students had made the estimates of the width of the lecture hall the room width was accurately measured and found to be 13.1 metres (43.0 feet). Use this additional information to determine which of the two types of estimates was more precise.

Ex. 2.2 For the mortality and water hardness data calculate the correlation between the two variables in each region, north and south.

Ex. 2.3 The standardised residuals calculated for the piston ring data are not entirely satisfactory for the reasons given in the text. An alternative residual suggested by Haberman (1973) is defined as the ratio of the standardised residuals and an adjustment:

$$\frac{\sqrt{(n_{jk} - E_{jk})^2 / E_{jk}}}{\sqrt{(1 - n_{j\cdot}/n)(1 - n_{\cdot k}/n)}}.$$

When the variables forming the contingency table are independent, the adjusted residuals are approximately normally distributed with mean zero and standard deviation one. Write a general R function to calculate both standardised and adjusted residuals for any $r \times c$ contingency table and apply it to the piston ring data.

Ex. 2.4 For the data in table `rearrests` estimate the difference between the probability of being rearrested after being tried in an adult court and in a juvenile court, and find a 95% confidence interval for the population difference.

CHAPTER 3

# Conditional Inference: Guessing Lengths, Suicides, Gastrointestinal Damage, and Newborn Infants

## 3.1 Introduction

There are many experimental designs or studies where the subjects are not a random sample from some well-defined population. For example, subjects recruited for a clinical trial are hardly ever a random sample from the set of all people suffering from a certain disease but are a selection of patients showing up for examination in a hospital participating in the trial. Usually, the subjects are randomly assigned to certain groups, for example a control and a treatment group, and the analysis needs to take this randomisation into account. In this chapter, we discuss such test procedures usually known as *(re)-randomisation* or *permutation tests*.

In the room width estimation experiment reported in Chapter 2 (see Table 2.1), 40 of the estimated widths (in feet) of 69 students and 26 of the estimated widths (in metres) of 44 students are tied. In fact, this violates one assumption of the *unconditional* test procedures applied in Chapter 2, namely that the measurements are drawn from a continuous distribution. In this chapter, the data will be reanalysed using conditional test procedures, i.e., statistical tests where the distribution of the test statistics under the null hypothesis is determined *conditionally* on the data at hand. A number of other data sets will also be considered in this chapter and these will now be described.

Mann (1981) reports a study carried out to investigate the causes of jeering or baiting behaviour by a crowd when a person is threatening to commit suicide by jumping from a high building. A hypothesis is that baiting is more likely to occur in warm weather. Mann (1981) classified 21 accounts of threatened suicide by two factors, the time of year and whether or not baiting occurred. The data are given in Table 3.1 and the question is whether they give any evidence to support the hypothesis? The data come from the northern hemisphere, so June–September are the warm months.

**Table 3.1**: `suicides` data. Crowd behaviour at threatened suicides.

| | Baiting | Nonbaiting |
|---|---|---|
| June–September | 8 | 4 |
| October–May | 2 | 7 |

*Source*: From Mann, L., *J. Pers. Soc. Psy.*, 41, 703–709, 1981. With permission.

The administration of non-steriodal anti-inflammatory drugs for patients suffering from arthritis induces gastrointestinal damage. Lanza (1987) and Lanza et al. (1988a,b, 1989) report the results of placebo-controlled randomised clinical trials investigating the prevention of gastrointestinal damage by the application of Misoprostol. The degree of the damage is determined by endoscopic examinations and the response variable is defined as the classification described in Table 3.2. Further details of the studies as well as the data can be found in Whitehead and Jones (1994). The data of the four studies are given in Tables 3.3, 3.4, 3.5 and 3.6.

**Table 3.2**: Classification system for the response variable.

| Classification | Endoscopy Examination |
|---|---|
| 1 | No visible lesions |
| 2 | One haemorrhage or erosion |
| 3 | 2-10 haemorrhages or erosions |
| 4 | 11-25 haemorrhages or erosions |
| 5 | More than 25 haemorrhages or erosions or an invasive ulcer of any size |

*Source*: From Whitehead, A. and Jones, N. M. B., *Stat. Med.*, 13, 2503–2515, 1994. With permission.

**Table 3.3**: `Lanza` data. Misoprostol randomised clinical trial from Lanza (1987).

| | | classification | | | | |
|---|---|---|---|---|---|---|
| `treatment` | | 1 | 2 | 3 | 4 | 5 |
| | Misoprostol | 21 | 2 | 4 | 2 | 0 |
| | Placebo | 2 | 2 | 4 | 9 | 13 |

**Table 3.4**: `Lanza` data. Misoprostol randomised clinical trial from Lanza et al. (1988a).

| treatment | classification 1 | 2 | 3 | 4 | 5 |
|---|---|---|---|---|---|
| Misoprostol | 20 | 4 | 6 | 0 | 0 |
| Placebo | 8 | 4 | 9 | 4 | 5 |

**Table 3.5**: `Lanza` data. Misoprostol randomised clinical trial from Lanza et al. (1988b).

| treatment | classification 1 | 2 | 3 | 4 | 5 |
|---|---|---|---|---|---|
| Misoprostol | 20 | 4 | 3 | 1 | 2 |
| Placebo | 0 | 2 | 5 | 5 | 17 |

**Table 3.6**: `Lanza` data. Misoprostol randomised clinical trial from Lanza et al. (1989).

| treatment | classification 1 | 2 | 3 | 4 | 5 |
|---|---|---|---|---|---|
| Misoprostol | 1 | 4 | 5 | 0 | 0 |
| Placebo | 0 | 0 | 0 | 4 | 6 |

Newborn infants exposed to antiepileptic drugs in utero have a higher risk of major and minor abnormalities of the face and digits. The inter-rater agreement in the assessment of babies with respect to the number of minor physical features was investigated by Carlin et al. (2000). In their paper, the agreement on total number of face anomalies for 395 newborn infants examined by a pediatrician and a research assistant is reported (see Table 3.7). One is interested in investigating whether the pediatrician and the research assistant agree above a chance level.

**Table 3.7**: `anomalies` data. Abnormalities of the face and digits of newborn infants exposed to antiepileptic drugs as assessed by a pediatrician (`MD`) and a research assistant (`RA`).

| | | RA | | | |
|---|---|---|---|---|---|
| MD | | 0 | 1 | 2 | 3 |
| | 0 | 235 | 41 | 20 | 2 |
| | 1 | 23 | 35 | 11 | 1 |
| | 2 | 3 | 8 | 11 | 3 |
| | 3 | 0 | 0 | 1 | 1 |

*Source*: From Carlin, J. B., et al., *Teratology*, 62, 406-412, 2000. With permission.

## 3.2 Conditional Test Procedures

The statistical test procedures applied in Chapter 2 all are defined for samples randomly drawn from a well-defined population. In many experiments however, this model is far from being realistic. For example in clinical trials, it is often impossible to draw a random sample from all patients suffering a certain disease. Commonly, volunteers and patients are recruited from hospital staff, relatives or people showing up for some examination. The test procedures applied in this chapter make no assumptions about random sampling or a specific model. Instead, the null distribution of the test statistics is computed conditionally on all random permutations of the data. Therefore, the procedures shown in the sequel are known as *permutation tests* or *(re)-randomisation tests*. For a general introduction we refer to the text books of Edgington (1987) and Pesarin (2001).

### *3.2.1 Testing Independence of Two Variables*

Based on $n$ pairs of measurements $(x_i, y_i)$ recorded for $n$ observational units we want to test the null hypothesis of the independence of $x$ and $y$. We may distinguish three situations: Both variables $x$ and $y$ are continuous, one is continuous and the other one is a factor or both $x$ and $y$ are factors. The special case of paired observations is treated in Section 3.2.2.

One class of test procedures for the above three situations are randomisation and permutation tests whose basic principles have been described by Fisher (1935) and Pitman (1937) and are best illustrated for the case of continuous measurements $y$ in two groups, i.e., the $x$ variable is a factor that can take values $x = 1$ or $x = 2$. The difference of the means of the $y$ values in both groups is an appropriate statistic for the assessment of the association of $y$

and $x$

$$T = \frac{\sum_{i=1}^{n} I(x_i = 1) y_i}{\sum_{i=1}^{n} I(x_i = 1)} - \frac{\sum_{i=1}^{n} I(x_i = 2) y_i}{\sum_{i=1}^{n} I(x_i = 2)}.$$

Here $I(x_i = 1)$ is the indication function which is equal to one if the condition $x_i = 1$ is true and zero otherwise. Clearly, under the null hypothesis of independence of $x$ and $y$ we expect the distribution of $T$ to be centred about zero.

Suppose that the group labels $x = 1$ or $x = 2$ have been assigned to the observational units by randomisation. When the result of the randomisation procedure is independent of the $y$ measurements, we are allowed to fix the $x$ values and shuffle the $y$ values randomly over and over again. Thus, we can compute, or at least approximate, the distribution of the test statistic $T$ under the conditions of the null hypothesis directly from the data $(x_i, y_i), i = 1, \dots, n$ by the so called *randomisation principle*. The test statistic $T$ is computed for a reasonable number of shuffled $y$ values and we can determine how many of the shuffled differences are at least as large as the test statistic $T$ obtained from the original data. If this proportion is small, smaller than $\alpha = 0.05$ say, we have good evidence that the assumption of independence of $x$ and $y$ is not realistic and we therefore can reject the null hypothesis. The proportion of larger differences is usually referred to as $p$-value.

A special approach is based on ranks assigned to the continuous $y$ values. When we replace the raw measurements $y_i$ by their corresponding ranks in the computation of $T$ and compare this test statistic with its null distribution we end up with the Wilcoxon Mann-Whitney rank sum test. The conditional distribution and the unconditional distribution of the Wilcoxon Mann-Whitney rank sum test as introduced in Chapter 2 coincide when the $y$ values are not tied. Without ties in the $y$ values, the ranks are simply the integers $1, 2, \dots, n$ and the unconditional (Chapter 2) and the conditonal view on the Wilcoxon Mann-Whitney test coincide.

In the case that both variables are nominal, the test statistic can be computed from the corresponding contingency table in which the observations $(x_i, y_i)$ are cross-classified. A general $r \times c$ contingency table may be written in the form of Table 2.6 where each cell $(j, k)$ is the number $n_{ij} = \sum_{i=1}^{n} I(x_i = j) I(y_i = k)$, see Chapter 2 for more details.

Under the null hypothesis of independence of $x$ and $y$, estimated expected values $E_{jk}$ for cell $(j, k)$ can be computed from the corresponding margin totals $E_{jk} = n_{j\cdot} n_{\cdot k} / n$ which are fixed for each randomisation of the data. The test statistic for assessing independence is

$$X^2 = \sum_{j=1}^{r} \sum_{k=1}^{c} \frac{(n_{jk} - E_{jk})^2}{E_{jk}}.$$

The exact distribution based on all permutations of the $y$ values for a similar test statistic can be computed by means of Fisher's exact test (Freeman and Halton, 1951). This test procedure is based on the hyper-geometric probability of the observed contingency table. All possible tables can be ordered with respect to this metric and $p$-values are computed from the fraction of tables more extreme than the observed one.

When both the $x$ and the $y$ measurements are numeric, the test statistic can be formulated as the product, i.e., by the sum of all $x_i y_i, i = 1, \ldots, n$. Again, we can fix the $x$ values and shuffle the $y$ values in order to approximate the distribution of the test statistic under the laws of the null hypothesis of independence of $x$ and $y$.

### *3.2.2 Testing Marginal Homogeneity*

In contrast to the independence problem treated above the data analyst is often confronted with situations where two (or more) measurements of one variable taken from the same observational unit are to be compared. In this case one assumes that the measurements are independent between observations and the test statistics are aggregated over all observations. Where two nominal variables are taken for each observation (for example see the case of McNemar's test for binary variables as discussed in Chapter 2), the measurement of each observation can be summarised by a $k \times k$ matrix with cell $(i, j)$ being equal to one if the first measurement is the $i$th level and the second measurement is the $j$th level. All other entries are zero. Under the null hypothesis of independence of the first and second measurement, all $k \times k$ matrices with exactly one non-zero element are equally likely. The test statistic is now based on the elementwise sum of all $n$ matrices.

## 3.3 Analysis Using R

### *3.3.1 Estimating the Width of a Room Revised*

The unconditional analysis of the room width estimated by two groups of students in Chapter 2 lead to the conclusion that the estimates in metres are slightly larger than the estimates in feet. Here, we reanalyse these data in a conditional framework. First, we convert metres into feet and store the vector of observations in a variable y:

```
R> data("roomwidth", package = "HSAUR")
R> convert <- ifelse(roomwidth$unit == "feet", 1, 3.28)
R> feet <- roomwidth$unit == "feet"
R> metre <- !feet
R> y <- roomwidth$width * convert
```

The test statistic is simply the difference in means

```
R> T <- mean(y[feet]) - mean(y[metre])
R> T

[1] -8.858893
```

```
R> hist(meandiffs)
R> abline(v = T, lty = 2)
R> abline(v = -T, lty = 2)
```

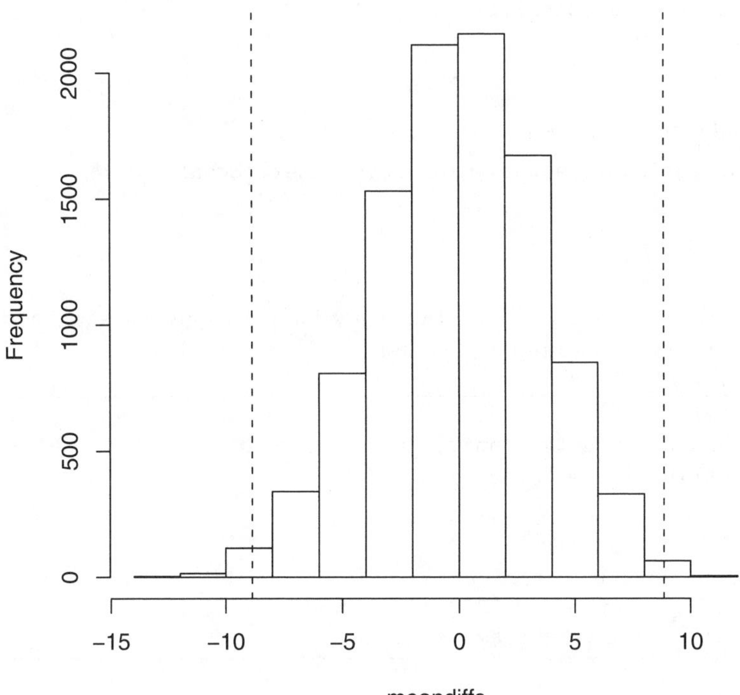


**Figure 3.1** Approximated conditional distribution of the difference of mean `roomwidth` estimates in the feet and metres group under the null hypothesis. The vertical lines show the negative and positive absolute value of the test statistic $T$ obtained from the original data.

In order to approximate the conditional distribution of the test statistic $T$ we compute 9999 test statistics for shuffled $y$ values. A permutation of the $y$ vector can be obtained from the `sample` function.

```
R> meandiffs <- double(9999)
R> for (i in 1:length(meandiffs)) {
+     sy <- sample(y)
+     meandiffs[i] <- mean(sy[feet]) - mean(sy[metre])
+ }
```

The distribution of the test statistic $T$ under the null hypothesis of independence of room width estimates and groups is depicted in Figure 3.1. Now, the value of the test statistic $T$ for the original unshuffled data can be compared with the distribution of $T$ under the null hypothesis (the vertical lines in Figure 3.1). The $p$-value, i.e., the proportion of test statistics $T$ larger than 8.859 or smaller than -8.859 is

```
R> greater <- abs(meandiffs) > abs(T)
R> mean(greater)

[1] 0.00790079
```

with a confidence interval of

```
R> binom.test(sum(greater), length(greater))$conf.int

[1] 0.006259976 0.009837149
attr(,"conf.level")
[1] 0.95
```

Note that the approximated conditional $p$-value is roughly the same as the $p$-value reported by the $t$-test in Chapter 2.

```
R> library("coin")
R> independence_test(y ~ unit, data = roomwidth,
+      distribution = "exact")

        Exact General Independence Test

data:  y by groups feet, metres
Z = -2.5491, p-value = 0.008492
alternative hypothesis: two.sided
```

**Figure 3.2** R output of the exact permutation test applied to the `roomwidth` data.

For some situations, including the analysis shown here, it is possible to compute the *exact* $p$-value, i.e., the $p$-value based on the distribution evaluated on all possible randomisations of the $y$ values. The function `independence_test` (package *coin*, Hothorn et al., 2005a, 2006a) can be used to compute the exact $p$-value as shown in Figure 3.2. Similarly, the exact conditional distribution of the Wilcoxon Mann-Whitney rank sum test can be computed by a function implemented in package *coin* as shown in Figure 3.3.

One should note that the $p$-values of the permutation test and the $t$-test coincide rather well and that the $p$-values of the Wilcoxon Mann-Whitney rank sum tests in their conditional and unconditional version are roughly three times as large due to the loss of information induced by only taking the ranking of the measurements into account. However, based on the results of the permutation test applied to the `roomwidth` data we can conclude that the estimates in metres are, on average, larger than the estimates in feet.

```
R> wilcox_test(y ~ unit, data = roomwidth,
+     distribution = "exact")

        Exact Wilcoxon Mann-Whitney Rank Sum Test

data:  y by groups feet, metres
Z = -2.1981, p-value = 0.02763
alternative hypothesis: true mu is not equal to 0
```

**Figure 3.3** R output of the exact conditional Wilcoxon rank sum test applied to the `roomwidth` data.

### *3.3.2 Crowds and Threatened Suicide*

The data in this case are in the form of a $2 \times 2$ contingency table and it might be thought that the chi-squared test could again be applied to test for the independence of crowd behaviour and time of year. However, the $\chi^2$-distribution as an approximation to the independence test statistic is bad when the expected frequencies are rather small. The problem is discussed in detail in Everitt (1992) and Agresti (1996). One solution is to use a conditional test procedure such as Fisher's exact test as described above. We can apply this test procedure using the R function `fisher.test` to the *table* `suicides` (see Figure 3.4).

```
R> data("suicides", package = "HSAUR")
R> fisher.test(suicides)

        Fisher's Exact Test for Count Data

data:  suicides
p-value = 0.0805
alternative hypothesis: true odds ratio is not equal to 1
95 percent confidence interval:
  0.7306872 91.0288231
sample estimates:
odds ratio
  6.302622
```

**Figure 3.4** R output of Fisher's exact test for the `suicides` data.

The resulting $p$-value obtained from the hypergeometric distribution is 0.08 (the asymptotic $p$-value associated with the $X^2$ statistic for this table is 0.115). There is no strong evidence of crowd behaviour being associated with time of year of threatened suicide, but the sample size is low and the test lacks power. Fisher's exact test can also be applied to larger than $2 \times 2$ tables, especially when there is concern that the cell frequencies are low (see Exercise 3.1).

### 3.3.3 Gastrointestinal Damages

Here we are interested in the comparison of two groups of patients, where one group received a placebo and the other one Misoprostol. In the trials shown here, the response variable is measured on an ordered scale – see Table 3.2. Data from four clinical studies are available and thus the observations are naturally grouped together. From the *data.frame* `Lanza` we can construct a three-way table as follows:

```
R> data("Lanza", package = "HSAUR")
R> xtabs(~treatment + classification + study, data = Lanza)
```

```
, , study = I

             classification
treatment      1  2  3  4  5
  Misoprostol 21  2  4  2  0
  Placebo      2  2  4  9 13

, , study = II

             classification
treatment      1  2  3  4  5
  Misoprostol 20  4  6  0  0
  Placebo      8  4  9  4  5

, , study = III

             classification
treatment      1  2  3  4  5
  Misoprostol 20  4  3  1  2
  Placebo      0  2  5  5 17

, , study = IV

             classification
treatment      1  2  3  4  5
  Misoprostol  1  4  5  0  0
  Placebo      0  0  0  4  6
```

We will first analyse each study separately and then show how one can investigate the effect of Misoprostol for all four studies simultaneously. Because the response is ordered, we take this information into account by assigning a score to each level of the response. Since the classifications are defined by the number of haemorrhages or erosions, the midpoint of the interval for each level is a reasonable choice, i.e., 0, 1, 6, 17 and 30 – compare those scores to the definitions given in Table 3.2. The corresponding linear-by-linear association tests extending the general Cochran-Mantel-Haenzel statistics (see Agresti, 2002, for further details) are implemented in package *coin*.

For the first study, the null hypothesis of independence of treatment and gastrointestinal damage, i.e., of no treatment effect of Misoprostol, is tested by

```
R> library("coin")
R> cmh_test(classification ~ treatment, data = Lanza,
+     scores = list(classification = c(0, 1, 6, 17,
+         30)), subset = Lanza$study == "I")

        Asymptotic Linear-by-Linear Association Test

data:  classification (ordered) by groups Misoprostol, Placebo
chi-squared = 28.8478, df = 1, p-value = 7.83e-08
```

and, by default, the conditional distribution is approximated by the corresponding limiting distribution. The $p$-value indicates a strong treatment effect. For the second study, the asymptotic $p$-value is a little bit larger

```
R> cmh_test(classification ~ treatment, data = Lanza,
+     scores = list(classification = c(0, 1, 6, 17,
+         30)), subset = Lanza$study == "II")

        Asymptotic Linear-by-Linear Association Test

data:  classification (ordered) by groups Misoprostol, Placebo
chi-squared = 12.0641, df = 1, p-value = 0.000514
```

and we make sure that the implied decision is correct by calculating a confidence interval for the exact $p$-value

```
R> p <- cmh_test(classification ~ treatment, data = Lanza,
+     scores = list(classification = c(0, 1, 6, 17,
+         30)), subset = Lanza$study == "II",
+     distribution = approximate(B = 19999))
R> pvalue(p)

[1] 5.00025e-05
99 percent confidence interval:
 2.506396e-07 3.714653e-04
```

The third and fourth study indicate a strong treatment effect as well

```
R> cmh_test(classification ~ treatment, data = Lanza,
+     scores = list(classification = c(0, 1, 6, 17,
+         30)), subset = Lanza$study == "III")

        Asymptotic Linear-by-Linear Association Test

data:  classification (ordered) by groups Misoprostol, Placebo
chi-squared = 28.1587, df = 1, p-value = 1.118e-07

R> cmh_test(classification ~ treatment, data = Lanza,
+     scores = list(classification = c(0, 1, 6, 17,
+         30)), subset = Lanza$study == "IV")
```

```
        Asymptotic Linear-by-Linear Association Test

data:  classification (ordered) by groups Misoprostol, Placebo
chi-squared = 15.7414, df = 1, p-value = 7.262e-05
```

At the end, a separate analysis for each study is unsatisfactory. Because the design of the four studies is the same, we can use **study** as a block variable and perform a global linear-association test investigating the treatment effect of Misoprostol in all four studies. The block variable can be incorporated into the *formula* by the | symbol.

```
R> cmh_test(classification ~ treatment | study, data = Lanza,
+     scores = list(classification = c(0, 1, 6, 17,
+         30)))

        Asymptotic Linear-by-Linear Association Test

data:  classification (ordered) by
         groups Misoprostol, Placebo
         stratified by study
chi-squared = 83.6188, df = 1, p-value < 2.2e-16
```

Based on this result, a strong treatment effect can be established.

### *3.3.4 Teratogenesis*

In this example, the medical doctor (MD) and the research assistant (RA) assessed the number of anomalies $(0, 1, 2$ or $3)$ for each of 395 babies:

```
R> anomalies <- as.table(matrix(c(235, 23, 3, 0, 41,
+     35, 8, 0, 20, 11, 11, 1, 2, 1, 3, 1), ncol = 4,
+     dimnames = list(MD = 0:3, RA = 0:3)))
R> anomalies

   RA
MD    0   1   2   3
  0 235  41  20   2
  1  23  35  11   1
  2   3   8  11   3
  3   0   0   1   1
```

We are interested in testing whether the number of anomalies assessed by the medical doctor differs structurally from the number reported by the research assistant. Because we compare *paired* observations, i.e., one pair of measurements for each newborn, a test of marginal homogeneity (a generalisation of McNemar's test, see Chapter 2) needs to be applied:

```
R> mh_test(anomalies)

        Asymptotic Marginal-Homogeneity Test

data:  response by
         groups MD, RA
         stratified by block
chi-squared = 21.2266, df = 3, p-value = 9.446e-05
```

The $p$-value indicates a deviation from the null hypothesis. However, the levels of the response are not treated as ordered. Similar to the analysis of the gastrointestinal damage data above, we can take this information into account by the definition of an appropriate score. Here, the number of anomalies is a natural choice:

```
R> mh_test(anomalies, scores = list(c(0, 1, 2, 3)))

        Asymptotic Marginal-Homogenity Test for Ordered Data

data:  response (ordered) by
         groups MD, RA
         stratified by block
chi-squared = 21.0199, df = 1, p-value = 4.545e-06
```

In our case, both versions coincide and one can conclude that the assessment of the number of anomalies differs between the medical doctor and the research assistant.

## 3.4 Summary

The analysis of randomised experiments, for example the analysis of randomised clinical trials such as the Misoprostol trial presented in this chapter, requires the application of conditional inferences procedures. In such experiments, the observations might not have been sampled from well-defined populations but are assigned to treatment groups, say, by a random procedure which is reiterated when randomisation tests are applied.

## Exercises

Ex. 3.1 Although in the past Fisher's test has been largely applied to sparse $2 \times 2$ tables, it can also be applied to larger tables, especially when there is concern about small values in some cells. Using the data displayed in Table 3.8 (taken from Mehta and Patel, 2003) which gives the distribution of the oral lesion site found in house-to-house surveys in three geographic regions of rural India, find the $p$-value from Fisher's test and the corresponding $p$-value from applying the usual chi-square test to the data. What are your conclusions?

**Table 3.8**: `orallesions` data. Oral lesions found in house-to-house surveys in three geographic regions of rural India.

| site of lesion | | region | |
|---|---|---|---|
| | Kerala | Gujarat | Andhra |
| Buccal mucosa | 8 | 1 | 8 |
| Commissure | 0 | 1 | 0 |
| Gingiva | 0 | 1 | 0 |
| Hard palate | 0 | 1 | 0 |
| Soft palate | 0 | 1 | 0 |
| Tongue | 0 | 1 | 0 |
| Floor of mouth | 1 | 0 | 1 |
| Alveolar ridge | 1 | 0 | 1 |

*Source*: From Mehta, C. and Patel, N., *StatXact-6: Statistical Software for Exact Nonparametric Inference*, Cytel Software Corporation, Cambridge, MA, 2003. With permission.

Ex. 3.2 Use the `mosaic` and `assoc` functions from the *vcd* package (Meyer et al., 2005) to create a graphical representation of the deviations from independence in the $2 \times 2$ contingency table shown in Table 3.1.

Ex. 3.3 Generate two groups with measurements following a normal distribution having different means. For multiple replications of this experiment (1000, say), compare the $p$-values of the Wilcoxon Mann-Whitney rank sum test and a permutation test (using `independence_test`). Where do the differences come from?

CHAPTER 4

# Analysis of Variance: Weight Gain, Foster Feeding in Rats, Water Hardness and Male Egyptian Skulls

## 4.1 Introduction

The data in Table 4.1 (from Hand et al., 1994) arise from an experiment to study the gain in weight of rats fed on four different diets, distinguished by amount of protein (low and high) and by source of protein (beef and cereal). Ten rats are randomised to each of the four treatments and the weight gain in grams recorded. The question of interest is how diet affects weight gain.

**Table 4.1**: `weightgain` data. Rat weight gain for diets differing by the amount of protein (`type`) and source of protein (`source`).

| source | type | weightgain | source | type | weightgain |
|---|---|---|---|---|---|
| Beef | Low | 90 | Cereal | Low | 107 |
| Beef | Low | 76 | Cereal | Low | 95 |
| Beef | Low | 90 | Cereal | Low | 97 |
| Beef | Low | 64 | Cereal | Low | 80 |
| Beef | Low | 86 | Cereal | Low | 98 |
| Beef | Low | 51 | Cereal | Low | 74 |
| Beef | Low | 72 | Cereal | Low | 74 |
| Beef | Low | 90 | Cereal | Low | 67 |
| Beef | Low | 95 | Cereal | Low | 89 |
| Beef | Low | 78 | Cereal | Low | 58 |
| Beef | High | 73 | Cereal | High | 98 |
| Beef | High | 102 | Cereal | High | 74 |
| Beef | High | 118 | Cereal | High | 56 |
| Beef | High | 104 | Cereal | High | 111 |
| Beef | High | 81 | Cereal | High | 95 |
| Beef | High | 107 | Cereal | High | 88 |
| Beef | High | 100 | Cereal | High | 82 |
| Beef | High | 87 | Cereal | High | 77 |
| Beef | High | 117 | Cereal | High | 86 |
| Beef | High | 111 | Cereal | High | 92 |

The data in Table 4.2 are from a foster feeding experiment with rat mothers and litters of four different genotypes: A, B, I and J (Hand et al., 1994). The measurement is the litter weight (in grams) after a trial feeding period. Here the investigator's interest lies in uncovering the effect of genotype of mother and litter on litter weight.

**Table 4.2**: `foster` data. Foster feeding experiment for rats with different genotypes of the litter (`litgen`) and mother (`motgen`).

| litgen | motgen | weight | litgen | motgen | weight |
|---|---|---|---|---|---|
| A | A | 61.5 | B | J | 40.5 |
| A | A | 68.2 | I | A | 37.0 |
| A | A | 64.0 | I | A | 36.3 |
| A | A | 65.0 | I | A | 68.0 |
| A | A | 59.7 | I | B | 56.3 |
| A | B | 55.0 | I | B | 69.8 |
| A | B | 42.0 | I | B | 67.0 |
| A | B | 60.2 | I | I | 39.7 |
| A | I | 52.5 | I | I | 46.0 |
| A | I | 61.8 | I | I | 61.3 |
| A | I | 49.5 | I | I | 55.3 |
| A | I | 52.7 | I | I | 55.7 |
| A | J | 42.0 | I | J | 50.0 |
| A | J | 54.0 | I | J | 43.8 |
| A | J | 61.0 | I | J | 54.5 |
| A | J | 48.2 | J | A | 59.0 |
| A | J | 39.6 | J | A | 57.4 |
| B | A | 60.3 | J | A | 54.0 |
| B | A | 51.7 | J | A | 47.0 |
| B | A | 49.3 | J | B | 59.5 |
| B | A | 48.0 | J | B | 52.8 |
| B | B | 50.8 | J | B | 56.0 |
| B | B | 64.7 | J | I | 45.2 |
| B | B | 61.7 | J | I | 57.0 |
| B | B | 64.0 | J | I | 61.4 |
| B | B | 62.0 | J | J | 44.8 |
| B | I | 56.5 | J | J | 51.5 |
| B | I | 59.0 | J | J | 53.0 |
| B | I | 47.2 | J | J | 42.0 |
| B | I | 53.0 | J | J | 54.0 |
| B | J | 51.3 | | | |

The data in Table 4.3 (from Hand et al., 1994) give four measurements made on Egyptian skulls from five epochs. The data has been collected with a view to deciding if there are any differences between the skulls from the five epochs. The measurements are:

`mb`: maximum breadths of the skull,

`bh`: basibregmatic heights of the skull,

`bl`: basialiveolar length of the skull, and

`nh`: nasal heights of the skull.

Non-constant measurements of the skulls over time would indicate interbreeding with immigrant populations.

**Table 4.3**: `skulls` data. Measurements of four variables taken from Egyptian skulls of five periods.

| epoch | mb | bh | bl | nh |
|---|---|---|---|---|
| c4000BC | 131 | 138 | 89 | 49 |
| c4000BC | 125 | 131 | 92 | 48 |
| c4000BC | 131 | 132 | 99 | 50 |
| c4000BC | 119 | 132 | 96 | 44 |
| c4000BC | 136 | 143 | 100 | 54 |
| c4000BC | 138 | 137 | 89 | 56 |
| c4000BC | 139 | 130 | 108 | 48 |
| c4000BC | 125 | 136 | 93 | 48 |
| c4000BC | 131 | 134 | 102 | 51 |
| c4000BC | 134 | 134 | 99 | 51 |
| c4000BC | 129 | 138 | 95 | 50 |
| c4000BC | 134 | 121 | 95 | 53 |
| c4000BC | 126 | 129 | 109 | 51 |
| c4000BC | 132 | 136 | 100 | 50 |
| c4000BC | 141 | 140 | 100 | 51 |
| c4000BC | 131 | 134 | 97 | 54 |
| c4000BC | 135 | 137 | 103 | 50 |
| c4000BC | 132 | 133 | 93 | 53 |
| c4000BC | 139 | 136 | 96 | 50 |
| c4000BC | 132 | 131 | 101 | 49 |
| c4000BC | 126 | 133 | 102 | 51 |
| c4000BC | 135 | 135 | 103 | 47 |
| c4000BC | 134 | 124 | 93 | 53 |
| c4000BC | 128 | 134 | 103 | 50 |
| c4000BC | 130 | 130 | 104 | 49 |
| ⋮ | ⋮ | ⋮ | ⋮ | ⋮ |

## 4.2 Analysis of Variance

For each of the data sets described in the previous section, the question of interest involves assessing whether certain populations differ in mean value for, in Tables 4.1 and 4.2, a single variable, and in Table 4.3, for a set of four variables. In the first two cases we shall use *analysis of variance* (ANOVA) and in the last *multivariate analysis of variance* (MANOVA) method for the analysis of this data.

Both Tables 4.1 and 4.2 are examples of *factorial designs*, with the factors in the first data set being amount of protein with two levels, and source of protein also with two levels. In the second the factors are the genotype of the mother and the genotype of the litter, both with four levels. The analysis of each data set can be based on the same model (see below) but the two data sets differ in that the first is *balanced*, i.e., there are the same number of observations in each cell, whereas the second is *unbalanced* having different numbers of observations in the 16 cells of the design. This distinction leads to complications in the analysis of the unbalanced design that we will come to in the next section. But the model used in the analysis of each is

$$y_{ijk} = \mu + \gamma_i + \beta_j + (\gamma\beta)_{ij} + \varepsilon_{ijk}$$

where $y_{ijk}$ represents the $k$th measurement made in cell $(i, j)$ of the factorial design, $\mu$ is the overall mean, $\gamma_i$ is the main effect of the first factor, $\beta_j$ is the main effect of the second factor, $(\gamma\beta)_{ij}$ is the interaction effect of the two factors and $\varepsilon_{ijk}$ is the residual or error term assumed to have a normal distribution with mean zero and variance $\sigma^2$. In R, the model is specified by a model *formula*. The *two-way layout with interactions* specified above reads

```
y ~ a + b + a:b
```

where the variable `a` is the first and the variable `b` is the second *factor*. The interaction term $(\gamma\beta)_{ij}$ is denoted by `a:b`. An equivalent model *formula* is

```
y ~ a * b
```

Note that the mean $\mu$ is implicitly defined in the *formula* shown above. In case $\mu = 0$, one needs to remove the intercept term from the *formula* explicitly, i.e.,

```
y ~ a + b + a:b - 1
```

For a more detailed description of model formulae we refer to R Development Core Team (2005a) and `help("lm")`.

The model as specified above is overparameterised, i.e., there are infinitively many solutions to the corresponding estimation equations, and so the parameters have to be constrained in some way, commonly by requiring them to sum to zero – see Everitt (2001) for a full discussion. The analysis of the rat weight gain data below explains some of these points in more detail (see also Chapter 5).

The model given above leads to a partition of the variation in the observations into parts due to main effects and interaction plus an error term that enables a series of $F$-tests to be calculated that can be used to test hypotheses about the main effects and the interaction. These calculations are generally set out in the familiar *analysis of variance table*. The assumptions made in deriving the $F$-tests are:

- The observations are independent of each other,
- The observations in each cell arise from a population having a normal distribution, and
- The observations in each cell are from populations having the same variance.

The multivariate analysis of variance, or MANOVA, is an extension of the univariate analysis of variance to the situation where a set of variables are measured on each individual or object observed. For the data in Table 4.3 there is a single factor, `epoch`, and four measurements taken on each skull; so we have a *one-way* MANOVA design. The linear model used in this case is

$$y_{ijh} = \mu_h + \gamma_{jh} + \varepsilon_{ijh}$$

where $\mu_h$ is the overall mean for variable $h$, $\gamma_{jh}$ is the effect of the $j$th level of the single factor on the $h$th variable, and $\varepsilon_{ijh}$ is a random error term. The vector $\varepsilon_{ij}^\top = (\varepsilon_{ij1}, \varepsilon_{ij2}, \dots, \varepsilon_{ijq})$ where $q$ is the number of response variables (four in the skull example) is assumed to have a multivariate normal distribution with null mean vector and covariance matrix, $\Sigma$, assumed to be the same in each level of the grouping factor. The hypothesis of interest is that the population mean vectors for the different levels of the grouping factor are the same.

In the multivariate situation, when there are more than two levels of the grouping factor, no single test statistic can be derived which is always the most powerful, for *all* types of departures from the null hypothesis of the equality of mean vector. A number of different test statistics are available which may give different results when applied to the same data set, although the final conclusion is often the same. The principal test statistics for the multivariate analysis of variance are *Hotelling-Lawley trace*, *Wilks' ratio of determinants*, *Roy's greatest root*, and the *Pillai trace*. Details are given in Morrison (2005).

## 4.3 Analysis Using R

### *4.3.1 Weight Gain in Rats*

Before applying analysis of variance to the data in Table 4.1 we should try to summarise the main features of the data by calculating means and standard deviations and by producing some hopefully informative graphs. The data is available in the *data.frame* `weightgain`. The following R code produces the required summary statistics

```
R> data("weightgain", package = "HSAUR")
R> tapply(weightgain$weightgain, list(weightgain$source,
+     weightgain$type), mean)
```

```
R> plot.design(weightgain)
```

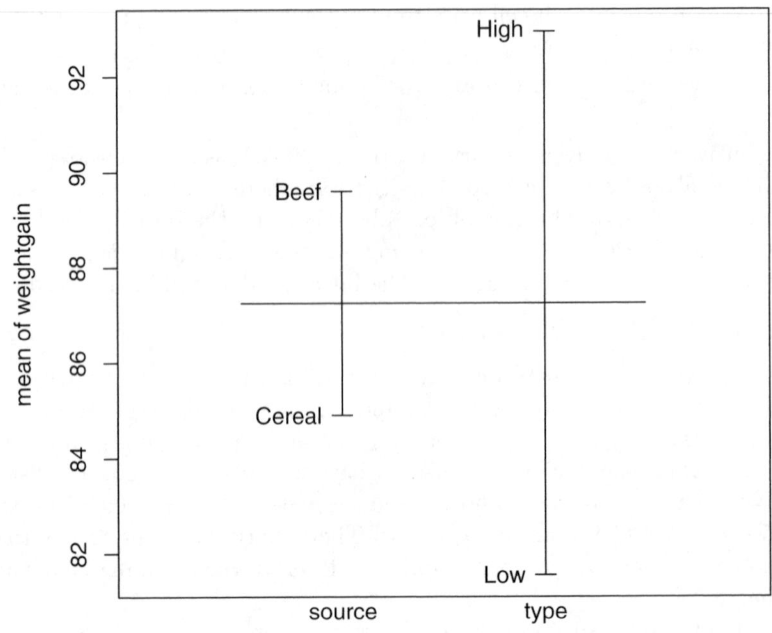


**Figure 4.1** Plot of mean weight gain for each level of the two factors.

```
         High  Low
Beef    100.0 79.2
Cereal   85.9 83.9
```

```
R> tapply(weightgain$weightgain, list(weightgain$source,
+     weightgain$type), sd)
```

```
           High      Low
Beef   15.13642 13.88684
Cereal 15.02184 15.70881
```

The cell variances are relatively similar and there is no apparent relationship between cell mean and cell variance so the homogeneity assumption of the analysis of variance looks like it is reasonable for these data. The plot of cell means in Figure 4.1 suggests that there is a considerable difference in weight gain for the amount of protein factor with the gain for the high-protein diet being far more than for the low-protein diet. A smaller difference is seen for the source factor with beef leading to a higher gain than cereal.

To apply analysis of variance to the data we can use the aov function in R and then the summary method to give us the usual analysis of variance table. The model *formula* specifies a two-way layout with interaction terms, where the first factor is source, and the second factor is type.

```
R> wg_aov <- aov(weightgain ~ source * type, data = weightgain)
```

```
R> summary(wg_aov)

            Df Sum Sq Mean Sq F value  Pr(>F)
source       1  220.9   220.9  0.9879 0.32688
type         1 1299.6  1299.6  5.8123 0.02114 *
source:type  1  883.6   883.6  3.9518 0.05447 .
Residuals   36 8049.4   223.6
---
Signif. codes:  0 '***' 0.001 '**' 0.01 '*' 0.05 '.' 0.1 ' ' 1
```

**Figure 4.2** R output of the ANOVA fit for the weightgain data.

The resulting analysis of variance table in Figure 4.2 shows that the main effect of type is highly significant confirming what was seen in Figure 4.1. The main effect of source is not significant. But interpretation of both these main effects is complicated by the type $\times$ source interaction which approaches significance at the 5% level. To try to understand this interaction effect it will be useful to plot the mean weight gain for low- and high-protein diets for each level of source of protein, beef and cereal. The required R code is given with Figure 4.3. From the resulting plot we see that for low-protein diets, the use of cereal as the source of the protein leads to a greater weight gain than using beef. For high-protein diets the reverse is the case with the beef/high diet leading to the highest weight gain.

The estimates of the intercept and the main and interaction effects can be extracted from the model fit by

```
R> coef(wg_aov)

          (Intercept)          sourceCereal               typeLow
                100.0                 -14.1                 -20.8
sourceCereal:typeLow
                 18.8
```

Note that the model was fitted with the restrictions $\gamma_1 = 0$ (corresponding to Beef) and $\beta_1 = 0$ (corresponding to High) because treatment contrasts were used as default as can be seen from

```
R> options("contrasts")

$contrasts
        unordered           ordered
"contr.treatment"      "contr.poly"
```

```
R> interaction.plot(weightgain$type, weightgain$source,
+     weightgain$weightgain)
```

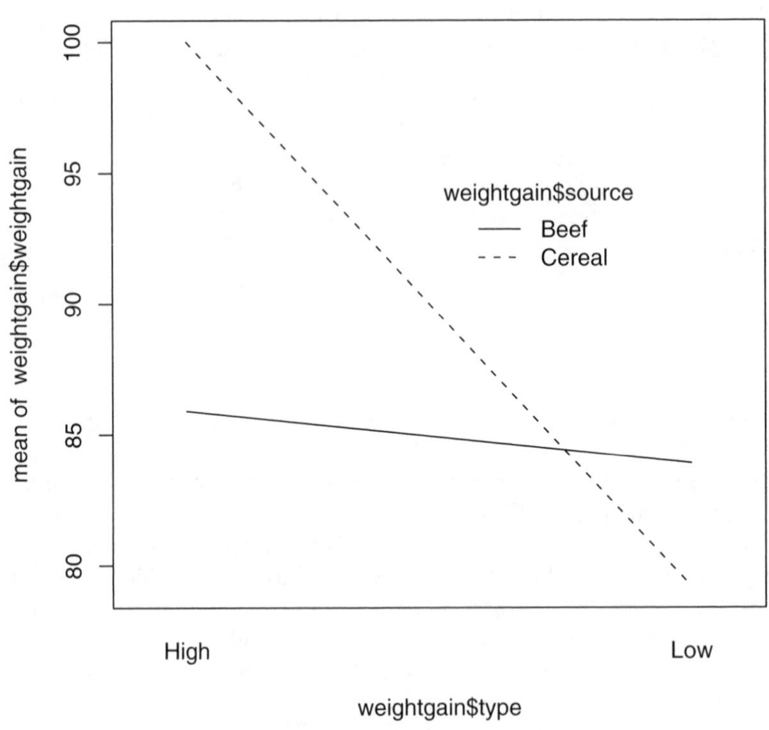


**Figure 4.3** Interaction plot of type × source.

Thus, the coefficient for `source` of $-14.1$ can be interpreted as an estimate of the difference $\gamma_2 - \gamma_1$. Alternatively, we can use the restriction $\sum_i \gamma_i = 0$ by

```
R> coef(aov(weightgain ~ source + type + source:type,
+     data = weightgain, contrasts = list(source = contr.sum)))

    (Intercept)           source1           typeLow
          92.95              7.05            -11.40
source1:typeLow
          -9.40
```

```
R> plot.design(foster)
```

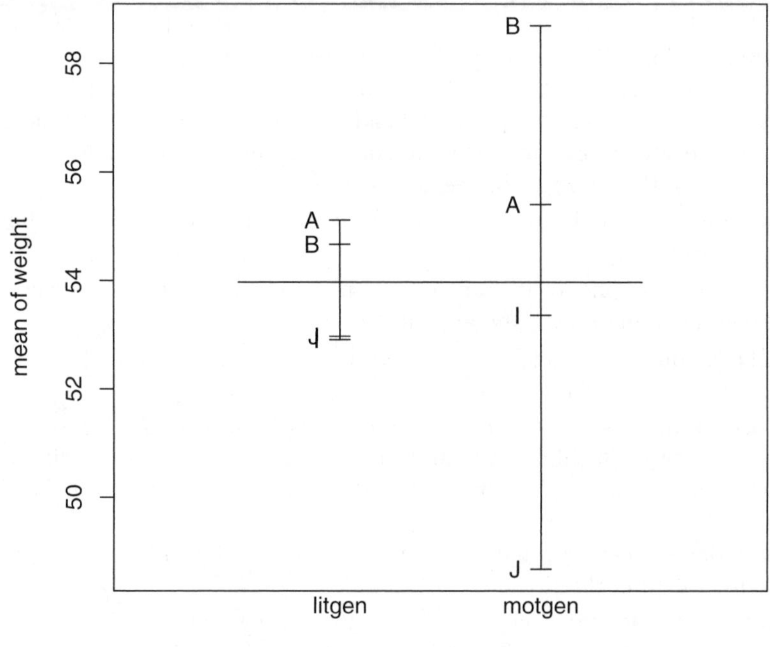


**Figure 4.4** Plot of mean litter weight for each level of the two factors for the `foster` data.

### *4.3.2 Foster Feeding of Rats of Different Genotype*

As in the previous subsection we will begin the analysis of the foster feeding data in Table 4.2 with a plot of the mean litter weight for the different genotypes of mother and litter (see Figure 4.4). The data are in the *data.frame* `foster`

```
R> data("foster", package = "HSAUR")
```

Figure 4.4 indicates that differences in litter weight for the four levels of mother's genotype are substantial; the corresponding differences for the genotype of the litter are much smaller.

As in the previous example we can now apply analysis of variance using the `aov` function, but there is a complication caused by the unbalanced nature of the data. Here where there are unequal numbers of observations in the 16 cells of the two-way layout, it is no longer possible to partition the variation in the data into *non-overlapping* or *orthogonal* sums of squares representing main effects and interactions. In an unbalanced two-way layout with factors $A$ and $B$ there is a proportion of the variance of the response variable that can be attributed to either $A$ or $B$. The consequence is that $A$ and $B$ together explain less of the variation of the dependent variable than the sum of which each explains alone. The result is that the sum of squares corresponding to a factor depends on which other terms are currently in the model for the observations, so the sums of squares depend on the order in which the factors are considered and represent a comparison of models. For example, for the order $a, b, a \times b$, the sums of squares are such that

- $SSa$: compares model containing only the $a$ main effect with one containing only the overall mean.
- $SSb|a$: compares model including both main effects, but no interaction, with one including only the main effect of $a$.
- $SSab|a, b$: compares model including an interaction and main effects with one including only main effects.

The use of these sums of squares (sometimes known as *Type I sums of squares*) in a series of tables in which the effects are considered in different orders provides the most appropriate approach to the analysis of unbalanced designs.

We can derive the two analyses of variance tables for the foster feeding example by applying the R code

```
R> summary(aov(weight ~ litgen * motgen, data = foster))
```

to give

```
               Df  Sum Sq Mean Sq F value   Pr(>F)   
litgen          3   60.16   20.05  0.3697 0.775221   
motgen          3  775.08  258.36  4.7632 0.005736 **
litgen:motgen   9  824.07   91.56  1.6881 0.120053   
Residuals      45 2440.82   54.24                    
---
Signif. codes:  0 '***' 0.001 '**' 0.01 '*' 0.05 '.' 0.1 ' ' 1
```

and then the code

```
R> summary(aov(weight ~ motgen * litgen, data = foster))
```

to give

```
               Df  Sum Sq Mean Sq F value   Pr(>F)   
motgen          3  771.61  257.20  4.7419 0.005869 **
litgen          3   63.63   21.21  0.3911 0.760004   
motgen:litgen   9  824.07   91.56  1.6881 0.120053   
Residuals      45 2440.82   54.24                    
---
Signif. codes:  0 '***' 0.001 '**' 0.01 '*' 0.05 '.' 0.1 ' ' 1
```

There are (small) differences in the sum of squares for the two main effects and, consequently, in the associated $F$-tests and $p$-values. This would not be true if in the previous example in Subsection 4.3.1 we had used the code

```
R> summary(aov(weightgain ~ type * source, data = weightgain))
```

instead of the code which produced Figure 4.2 (readers should confirm that this is the case).

Although for the foster feeding data the differences in the two analyses of variance with different orders of main effects are very small, this may not always be the case and care is needed in dealing with unbalanced designs. For a more complete discussion see Nelder (1977) and Aitkin (1978).

Both ANOVA tables indicate that the main effect of mother's genotype is highly significant and that genotype B leads to the greatest litter weight and genotype J to the smallest litter weight.

We can investigate the effect of genotype B on litter weight in more detail by the use of *multiple comparison procedures* (see Everitt, 1996). Such procedures allow a comparison of all pairs of levels of a factor whilst maintaining the nominal significance level at its selected value and producing adjusted confidence intervals for mean differences. One such procedure is called *Tukey honest significant differences* suggested by Tukey (1953), see Hochberg and Tamhane (1987) also. Here, we are interested in simultaneous confidence intervals for the weight differences between all four genotypes of the mother. First, an ANOVA model is fitted

```
R> foster_aov <- aov(weight ~ litgen * motgen, data = foster)
```

which serves as the basis of the multiple comparisons, here with allpair differences by

```
R> foster_hsd <- TukeyHSD(foster_aov, "motgen")
R> foster_hsd
```

```
  Tukey multiple comparisons of means
    95% family-wise confidence level

Fit: aov(formula = weight ~ litgen * motgen, data = foster)

$motgen
          diff        lwr        upr     p adj
B-A  3.330369  -3.859729 10.5204672 0.6078581
I-A -1.895574  -8.841869  5.0507207 0.8853702
J-A -6.566168 -13.627285  0.4949498 0.0767540
I-B -5.225943 -12.416041  1.9641552 0.2266493
J-B -9.896537 -17.197624 -2.5954489 0.0040509
J-I -4.670593 -11.731711  2.3905240 0.3035490
```

A convenient `plot` method exists for this object and we can get a graphical representation of the multiple confidence intervals as shown in Figure 4.5. It appears that there is only evidence for a difference in the B and J genotypes.

### *4.3.3 Water Hardness and Mortality*

The water hardness and mortality data for 61 large towns in England and Wales (see Table 2.3) was analysed in Chapter 2 and here we will extend the

```
R> plot(foster_hsd)
```

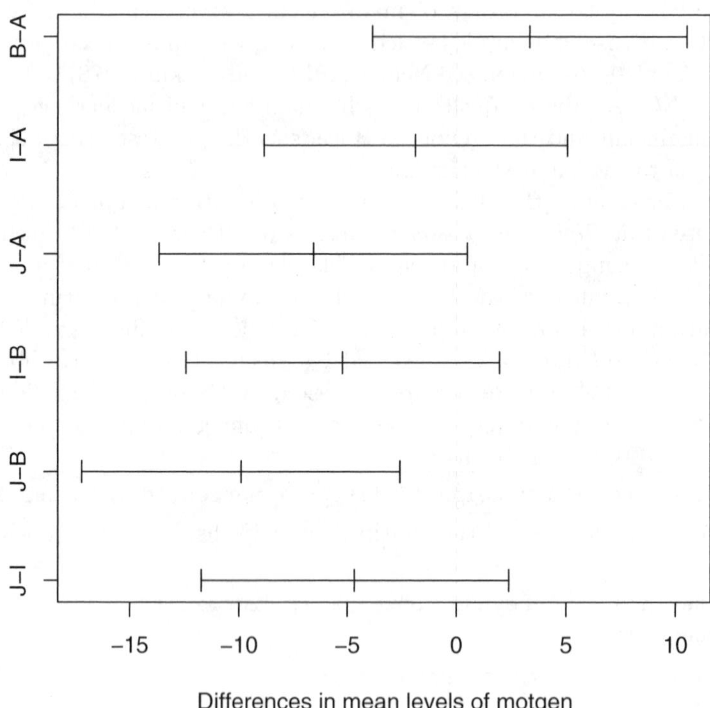


**Figure 4.5** Graphical presentation of multiple comparison results for the `foster` feeding data.

analysis by an assessment of the differences of both hardness and mortality in the North or South. The hypothesis that the two-dimensional mean-vector of water hardness and mortality is the same for cities in the North and the South can be tested by *Hotelling-Lawley* test in a multivariate analysis of variance framework. The R function `manova` can be used to fit such a model and the corresponding `summary` method performs the test specified by the `test` argument

```
R> data("water", package = "HSAUR")
R> summary(manova(cbind(hardness, mortality) ~ location,
+     data = water), test = "Hotelling-Lawley")

          Df Hotelling-Lawley approx F num Df den Df    Pr(>F)
location   1           0.9002  26.1062      2     58 8.217e-09
```

```
Residuals 59

location   ***
Residuals
---
Signif. codes:  0 '***' 0.001 '**' 0.01 '*' 0.05 '.' 0.1 ' ' 1
```

The `cbind` statement in the left hand side of the formula indicates that a *multivariate* response variable is to be modelled. The $p$-value associated with the *Hotelling-Lawley* statistic is very small and there is strong evidence that the mean vectors of the two variables are not the same in the two regions. Looking at the sample means

```
R> tapply(water$hardness, water$location, mean)

   North    South
30.40000 69.76923

R> tapply(water$mortality, water$location, mean)

   North    South
1633.600 1376.808
```

we see large differences in the two regions both in water hardness and mortality, where low mortality is associated with hard water in the South and high mortality with soft water in the North (see Figure 2.8 also).

### *4.3.4 Male Egyptian Skulls*

We can begin by looking at a table of mean values for the four measurements within each of the five epochs. The measurements are available in the *data.frame* `skulls` and we can compute the means over all epochs by

```
R> data("skulls", package = "HSAUR")
R> means <- aggregate(skulls[, c("mb", "bh", "bl",
+     "nh")], list(epoch = skulls$epoch), mean)
R> means

    epoch       mb       bh       bl       nh
1 c4000BC 131.3667 133.6000 99.16667 50.53333
2 c3300BC 132.3667 132.7000 99.06667 50.23333
3 c1850BC 134.4667 133.8000 96.03333 50.56667
4  c200BC 135.5000 132.3000 94.53333 51.96667
5  cAD150 136.1667 130.3333 93.50000 51.36667
```

It may also be useful to look at these means graphically and this could be done in a variety of ways. Here we construct a scatterplot matrix of the means using the code attached to Figure 4.6.

There appear to be quite large differences between the epoch means, at least on some of the four measurements. We can now test for a difference more formally by using MANOVA with the following R code to apply each of the four possible test criteria mentioned earlier;

```
R> pairs(means[, -1], panel = function(x, y) {
+     text(x, y, abbreviate(levels(skulls$epoch)))
+ })
```

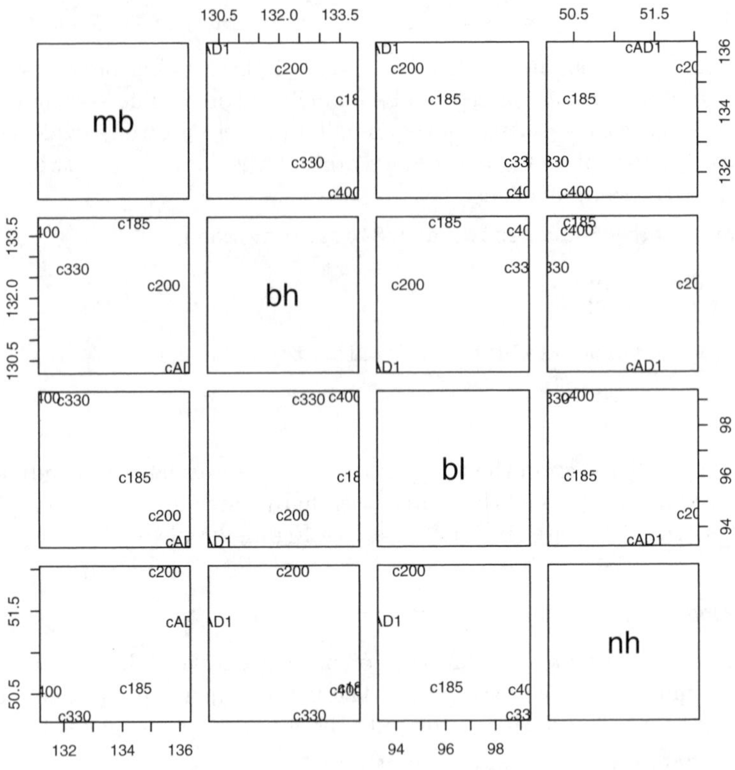


**Figure 4.6** Scatterplot matrix of epoch means for Egyptian **skulls** data.

```
R> skulls_manova <- manova(cbind(mb, bh, bl, nh) ~
+     epoch, data = skulls)
R> summary(skulls_manova, test = "Pillai")

           Df Pillai approx F num Df den Df    Pr(>F)
epoch       4 0.3533   3.5120     16    580 4.675e-06 ***
Residuals 145
---
Signif. codes:  0 '***' 0.001 '**' 0.01 '*' 0.05 '.' 0.1 ' ' 1

R> summary(skulls_manova, test = "Wilks")

             Df  Wilks approx F num Df den Df   Pr(>F)
epoch      4.00 0.6636   3.9009  16.00 434.45 7.01e-07 ***
```

```
Residuals 145.00
---
Signif. codes:  0 '***' 0.001 '**' 0.01 '*' 0.05 '.' 0.1 ' ' 1
```

```
R> summary(skulls_manova, test = "Hotelling-Lawley")
```

```
           Df Hotelling-Lawley approx F num Df den Df
epoch       4           0.4818   4.2310     16    562
Residuals 145
             Pr(>F)
epoch     8.278e-08 ***
Residuals
---
Signif. codes:  0 '***' 0.001 '**' 0.01 '*' 0.05 '.' 0.1 ' ' 1
```

```
R> summary(skulls_manova, test = "Roy")
```

```
           Df    Roy approx F num Df den Df    Pr(>F)
epoch       4 0.4251  15.4097      4    145 1.588e-10 ***
Residuals 145
---
Signif. codes:  0 '***' 0.001 '**' 0.01 '*' 0.05 '.' 0.1 ' ' 1
```

The $p$-value associated with each four test criteria is very small and there is strong evidence that the skull measurements differ between the five epochs. We might now move on to investigate which epochs differ and on which variables. We can look at the univariate $F$-tests for each of the four variables by using the code

```
R> summary.aov(manova(cbind(mb, bh, bl, nh) ~ epoch,
+      data = skulls))
```

```
 Response mb :
             Df  Sum Sq Mean Sq F value    Pr(>F)
epoch         4  502.83  125.71  5.9546 0.0001826 ***
Residuals   145 3061.07   21.11
---
Signif. codes:  0 '***' 0.001 '**' 0.01 '*' 0.05 '.' 0.1 ' ' 1

 Response bh :
             Df Sum Sq Mean Sq F value  Pr(>F)
epoch         4  229.9    57.5  2.4474 0.04897 *
Residuals   145 3405.3    23.5
---
Signif. codes:  0 '***' 0.001 '**' 0.01 '*' 0.05 '.' 0.1 ' ' 1

 Response bl :
             Df Sum Sq Mean Sq F value    Pr(>F)
epoch         4  803.3   200.8  8.3057 4.636e-06 ***
Residuals   145 3506.0    24.2
---
Signif. codes:  0 '***' 0.001 '**' 0.01 '*' 0.05 '.' 0.1 ' ' 1
```

```
 Response nh :
              Df  Sum Sq Mean Sq F value Pr(>F)
epoch          4   61.20   15.30   1.507 0.2032
Residuals    145 1472.13   10.15
```

We see that the results for the maximum breadths (`mb`) and basialiveolar length (`bl`) are highly significant, with those for the other two variables, in particular for nasal heights (`nh`), suggesting little evidence of a difference. To look at the pairwise multivariate tests (any of the four test criteria are equivalent in the case of a one-way layout with two levels only) we can use the `summary` method and `manova` function as follows:

```
R> summary(manova(cbind(mb, bh, bl, nh) ~ epoch, data = skulls,
+      subset = epoch %in% c("c4000BC", "c3300BC")))

          Df  Pillai approx F num Df den Df Pr(>F)
epoch      1 0.02767  0.39135      4     55  0.814
Residuals 58

R> summary(manova(cbind(mb, bh, bl, nh) ~ epoch, data = skulls,
+      subset = epoch %in% c("c4000BC", "c1850BC")))

          Df Pillai approx F num Df den Df  Pr(>F)
epoch      1 0.1876   3.1744      4     55 0.02035 *
Residuals 58
---
Signif. codes:  0 '***' 0.001 '**' 0.01 '*' 0.05 '.' 0.1 ' ' 1

R> summary(manova(cbind(mb, bh, bl, nh) ~ epoch, data = skulls,
+      subset = epoch %in% c("c4000BC", "c200BC")))

          Df Pillai approx F num Df den Df    Pr(>F)
epoch      1 0.3030   5.9766      4     55 0.0004564 ***
Residuals 58
---
Signif. codes:  0 '***' 0.001 '**' 0.01 '*' 0.05 '.' 0.1 ' ' 1

R> summary(manova(cbind(mb, bh, bl, nh) ~ epoch, data = skulls,
+      subset = epoch %in% c("c4000BC", "cAD150")))

          Df Pillai approx F num Df den Df    Pr(>F)
epoch      1 0.3618   7.7956      4     55 4.736e-05 ***
Residuals 58
---
Signif. codes:  0 '***' 0.001 '**' 0.01 '*' 0.05 '.' 0.1 ' ' 1
```

To keep the overall significance level for the set of all pairwise multivariate tests under some control (and still maintain a reasonable power), Stevens (2001) recommends setting the nominal level $\alpha = 0.15$ and carrying out each test at the $\alpha/m$ level where $m$ s the number of tests performed. The results of the four pairwise tests suggest that as the epochs become further separated in time the four skull measurements become increasingly distinct.

For more details of applying multiple comparisons in the multivariate situation see Stevens (2001).

## 4.4 Summary

Analysis of variance is one of the most widely used of statistical techniques and is easily applied using R as is the extension to multivariate data. An analysis of variance needs to be supplemented by graphical material prior to formal analysis and often to more detailed investigation of group differences using multiple comparison techniques.

### Exercises

Ex. 4.1 Examine the residuals (observed value – fitted value) from fitting a main effects only model to the data in Table 4.1. What conclusions do you draw?

Ex. 4.2 The data in Table 4.4 below arise from a sociological study of Australian Aboriginal and white children reported by Quine (1975). In this study, children of both sexes from four age groups (final grade in primary schools and first, second and third form in secondary school) and from two cultural groups were used. The children in age group were classified as slow or average learners. The response variable was the number of days absent from school during the school year. (Children who had suffered a serious illness during the years were excluded.) Carry out what you consider to be an appropriate analysis of variance of the data noting that (i) there are unequal numbers of observations in each cell and (ii) the response variable here is a count. Interpret your results with the aid of some suitable tables of means and some informative graphs.

**Table 4.4**: `schooldays` data. Days absent from school.

| race | sex | school | learner | absent |
|---:|---|---:|---:|---:|
| aboriginal | male | F0 | slow | 2 |
| aboriginal | male | F0 | slow | 11 |
| aboriginal | male | F0 | slow | 14 |
| aboriginal | male | F0 | average | 5 |
| aboriginal | male | F0 | average | 5 |
| aboriginal | male | F0 | average | 13 |
| aboriginal | male | F0 | average | 20 |
| aboriginal | male | F0 | average | 22 |
| aboriginal | male | F1 | slow | 6 |
| aboriginal | male | F1 | slow | 6 |
| aboriginal | male | F1 | slow | 15 |
| aboriginal | male | F1 | average | 7 |
| aboriginal | male | F1 | average | 14 |
| aboriginal | male | F2 | slow | 6 |
| aboriginal | male | F2 | slow | 32 |
| ⋮ | ⋮ | ⋮ | ⋮ | ⋮ |

Ex. 4.3 The data in Table 4.5 arise from a large study of risk taking (see Timm, 2002). Students were randomly assigned to three different treatments labelled AA, C and NC. Students were administered two parallel forms of a test called 'low' and 'high'. Carry out a test of the equality of the bivariate means of each treatment population.

**Table 4.5**: `students` data. Treatment and results of two tests in three groups of students.

| treatment | low | high | treatment | low | high |
|---|---|---|---|---|---|
| AA | 8 | 28 | C | 34 | 4 |
| AA | 18 | 28 | C | 34 | 4 |
| AA | 8 | 23 | C | 44 | 7 |
| AA | 12 | 20 | C | 39 | 5 |
| AA | 15 | 30 | C | 20 | 0 |
| AA | 12 | 32 | C | 43 | 11 |
| AA | 18 | 31 | NC | 50 | 5 |
| AA | 29 | 25 | NC | 57 | 51 |
| AA | 6 | 28 | NC | 62 | 52 |
| AA | 7 | 28 | NC | 56 | 52 |
| AA | 6 | 24 | NC | 59 | 40 |
| AA | 14 | 30 | NC | 61 | 68 |
| AA | 11 | 23 | NC | 66 | 49 |
| AA | 12 | 20 | NC | 57 | 49 |
| C | 46 | 13 | NC | 62 | 58 |
| C | 26 | 10 | NC | 47 | 58 |
| C | 47 | 22 | NC | 53 | 40 |
| C | 44 | 14 | | | |

*Source*: From Timm, N. H., *Applied Multivariate Analysis*, Springer, New York, 2002. With kind permission of Springer Science and Business Media.

CHAPTER 5

# Multiple Linear Regression: Cloud Seeding

## 5.1 Introduction

Weather modification, or cloud seeding, is the treatment of individual clouds or storm systems with various inorganic and organic materials in the hope of achieving an increase in rainfall. Introduction of such material into a cloud that contains supercooled water, that is, liquid water colder than zero degrees of Celsius, has the aim of inducing freezing, with the consequent ice particles growing at the expense of liquid droplets and becoming heavy enough to fall as rain from clouds that otherwise would produce none. The data shown in Table 5.1 were collected in the summer of 1975 from an experiment to investigate the use of massive amounts of silver iodide (100 to 1000 grams per cloud) in cloud seeding to increase rainfall (Woodley et al., 1977). In the experiment, which was conducted in an area of Florida, 24 days were judged suitable for seeding on the basis that a measured suitability criterion, denoted *S-Ne*, was not less than 1.5. Here *S* is the 'seedability', the difference between the maximum height of a cloud if seeded and the same cloud if not seeded predicted by a suitable cloud model, and *Ne* is the number of hours between 1300 and 1600 G.M.T. with 10 centimetre echoes in the target; this quantity biases the decision for experimentation against naturally rainy days. Consequently, optimal days for seeding are those on which seedability is large and the natural rainfall early in the day is small.

On suitable days, a decision was taken at random as to whether to seed or not. For each day the following variables were measured:

`seeding`: a factor indicating whether seeding action occured (yes or no),

`time`: number of days after the first day of the experiment,

`cloudcover`: the percentage cloud cover in the experimental area, measured using radar,

`prewetness`: the total rainfall in the target area one hour before seeding (in cubic metres $\times 10^7$),

`echomotion`: a factor showing whether the radar echo was moving or stationary,

`rainfall`: the amount of rain in cubic metres $\times 10^7$,

`sne`: suitability criterion, see above.

The objective in analysing these data is to see how rainfall is related to the explanatory variables and, in particular, to determine the effectiveness of seeding. The method to be used is *multiple linear regression*.

**Table 5.1**: `clouds` data. Cloud seeding experiments in Florida – see above for explanations of the variables.

| seeding | time | sne | cloudcover | prewetness | echomotion | rainfall |
|---|---|---|---|---|---|---|
| no | 0 | 1.75 | 13.4 | 0.274 | stationary | 12.85 |
| yes | 1 | 2.70 | 37.9 | 1.267 | moving | 5.52 |
| yes | 3 | 4.10 | 3.9 | 0.198 | stationary | 6.29 |
| no | 4 | 2.35 | 5.3 | 0.526 | moving | 6.11 |
| yes | 6 | 4.25 | 7.1 | 0.250 | moving | 2.45 |
| no | 9 | 1.60 | 6.9 | 0.018 | stationary | 3.61 |
| no | 18 | 1.30 | 4.6 | 0.307 | moving | 0.47 |
| no | 25 | 3.35 | 4.9 | 0.194 | moving | 4.56 |
| no | 27 | 2.85 | 12.1 | 0.751 | moving | 6.35 |
| yes | 28 | 2.20 | 5.2 | 0.084 | moving | 5.06 |
| yes | 29 | 4.40 | 4.1 | 0.236 | moving | 2.76 |
| yes | 32 | 3.10 | 2.8 | 0.214 | moving | 4.05 |
| no | 33 | 3.95 | 6.8 | 0.796 | moving | 5.74 |
| yes | 35 | 2.90 | 3.0 | 0.124 | moving | 4.84 |
| yes | 38 | 2.05 | 7.0 | 0.144 | moving | 11.86 |
| no | 39 | 4.00 | 11.3 | 0.398 | moving | 4.45 |
| no | 53 | 3.35 | 4.2 | 0.237 | stationary | 3.66 |
| yes | 55 | 3.70 | 3.3 | 0.960 | moving | 4.22 |
| no | 56 | 3.80 | 2.2 | 0.230 | moving | 1.16 |
| yes | 59 | 3.40 | 6.5 | 0.142 | stationary | 5.45 |
| yes | 65 | 3.15 | 3.1 | 0.073 | moving | 2.02 |
| no | 68 | 3.15 | 2.6 | 0.136 | moving | 0.82 |
| yes | 82 | 4.01 | 8.3 | 0.123 | moving | 1.09 |
| no | 83 | 4.65 | 7.4 | 0.168 | moving | 0.28 |

## 5.2 Multiple Linear Regression

Assume $y_i$ represents the value of the response variable on the $i$th individual, and that $x_{i1}, x_{i2}, \dots, x_{iq}$ represents the individual's values on $q$ explanatory variables, with $i = 1, \dots, n$. The multiple linear regression model is given by

$$y_i = \beta_0 + \beta_1 x_{i1} + \ldots + \beta_q x_{iq} + \varepsilon_i.$$

The residual or error terms $\varepsilon_i$, $i = 1, \dots, n$, are assumed to be independent random variables having a normal distribution with mean zero and constant variance $\sigma^2$. Consequently, the distribution of the random response variable,

$y$, is also normal with expected value given by the linear combination of the explanatory variables

$$\mathsf{E}(y|x_1, \dots, x_q) = \beta_0 + \beta_1 x_1 + \dots + \beta_q x_q$$

and with variance $\sigma^2$.

The parameters of the model $\beta_k$, $k = 1, \dots, q$, are known as regression coefficients with $\beta_0$ corresponding to the overall mean. They represent the expected change in the response variable associated with a unit change in the corresponding explanatory variable, when the remaining explanatory variables are held constant. The *linear* in multiple linear regression applies to the regression parameters, not to the response or explanatory variables. Consequently, models in which, for example, the logarithm of a response variable is modelled in terms of quadratic functions of some of the explanatory variables would be included in this class of models.

The multiple linear regression model can be written most conveniently for all $n$ individuals by using matrices and vectors as $\mathbf{y} = \mathbf{X}\beta + \varepsilon$ where $\mathbf{y}^\top = (y_1, \dots, y_n)$ is the vector of response variables, $\beta^\top = (\beta_0, \beta_1, \dots, \beta_q)$ is the vector of regression coefficients, and $\varepsilon^\top = (\varepsilon_1, \dots, \varepsilon_n)$ are the error terms. The *design* or *model matrix* $\mathbf{X}$ consists of the $q$ continuously measured explanatory variables and a column of ones corresponding to the *intercept* term

$$\mathbf{X} = \begin{pmatrix} 1 & x_{11} & x_{12} & \dots & x_{1q} \\ 1 & x_{21} & x_{22} & \dots & x_{2q} \\ \vdots & \vdots & \vdots & \ddots & \vdots \\ 1 & x_{n1} & x_{n2} & \dots & x_{nq} \end{pmatrix}.$$

In case one or more of the explanatory variables are nominal or ordinal variables, they are represented by a zero-one dummy coding. Assume that $x_1$ is a factor at $k$ levels, the submatrix of $\mathbf{X}$ corresponding to $x_1$ is a $n \times k$ matrix of zeros and ones, where the $j$th element in the $i$th row is one when $x_{i1}$ is at the $j$th level.

Assuming that the cross-product $\mathbf{X}^\top\mathbf{X}$ is non-singular, i.e., can be inverted, then the least squares estimator of the parameter vector $\beta$ is unique and can be calculated by $\hat{\beta} = (\mathbf{X}^\top\mathbf{X})^{-1}\mathbf{X}^\top\mathbf{y}$. The expectation and covariance of this estimator $\hat{\beta}$ are given by $\mathsf{E}(\hat{\beta}) = \beta$ and $\mathsf{Var}(\hat{\beta}) = \sigma^2(\mathbf{X}^\top\mathbf{X})^{-1}$. The diagonal elements of the covariance matrix $\mathsf{Var}(\hat{\beta})$ give the variances of $\hat{\beta}_j, j = 0, \dots, q$, whereas the off diagonal elements give the covariances between pairs of $\hat{\beta}_j$ and $\hat{\beta}_k$. The square roots of the diagonal elements of the covariance matrix are thus the standard errors of the estimates $\hat{\beta}_j$.

If the cross-product $\mathbf{X}^\top\mathbf{X}$ is singular we need to reformulate the model to $\mathbf{y} = \mathbf{X}\mathbf{C}\beta^\star + \varepsilon$ such that $\mathbf{X}^\star = \mathbf{X}\mathbf{C}$ has full rank. The matrix $\mathbf{C}$ is called *contrast matrix* in S and R and the result of the model fit is an estimate $\hat{\beta}^\star$. For the theoretical details we refer to Searle (1971), the implementation of contrasts in S and R is discussed by Chambers and Hastie (1992) and Venables and Ripley (2002).

The regression analysis can be assessed using the following analysis of variance table (Table 5.2):

**Table 5.2**: Analysis of variance table for the multiple linear regression model.

| Source of variation | Sum of squares | Degrees of freedom |
|---|---|---|
| Regression | $\sum_{i=1}^{n}(\hat{y}_i - \bar{y})^2$ | $q$ |
| Residual | $\sum_{i=1}^{n}(\hat{y}_i - y_i)^2$ | $n - q - 1$ |
| Total | $\sum_{i=1}^{n}(y_i - \bar{y})^2$ | $n - 1$ |

where $\hat{y}_i$ is the predicted value of the response variable for the $i$th individual $\hat{y}_i = \hat{\beta}_0 + \hat{\beta}_1 x_{i1} + \ldots + \hat{\beta}_q x_{q1}$ and $\bar{y} = \sum_{i=1}^{n} y_i/n$ is the mean of the response variable.

The mean square ratio

$$F = \frac{\sum_{i=1}^{n}(\hat{y}_i - \bar{y})^2/q}{\sum_{i=1}^{n}(\hat{y}_i - y_i)^2/(n - q - 1)}$$

provides an $F$-test of the general hypothesis

$$H_0 : \beta_1 = \ldots = \beta_q = 0.$$

Under $H_0$, the test statistic $F$ has an $F$-distribution with $q$ and $n - q - 1$ degrees of freedom. An estimate of the variance $\sigma^2$ is

$$\hat{\sigma}^2 = \frac{1}{n - q - 1}\sum_{i=1}^{n}(y_i - \hat{y}_i)^2.$$

The correlation between the observed values $y_i$ and the fitted values $\hat{y}_i$ is known as the multiple correlation coefficient. Individual regression coefficients can be assessed by using the ratio $t$-statistics $t_j = \hat{\beta}_j/\sqrt{\mathsf{Var}(\hat{\beta})_{jj}}$, although these ratios should only be used as rough guides to the 'significance' of the coefficients. The problem of selecting the 'best' subset of variables to be included in a model is one of the most delicate ones in statistics and we refer to Miller (2002) for the theoretical details and practical limitations (and see Exercise 5.4).

## 5.3 Analysis Using R

Prior to applying multiple regression to the data it will be useful to look at some graphics to assess their major features. Here we will construct boxplots of the rainfall in each category of the dichotomous explanatory variables and scatterplots of rainfall against each of the continuous explanatory variables.

Both the boxplots (Figure 5.1) and the scatterplots (Figure 5.2) show some evidence of outliers. The row names of the extreme observations in the `clouds` *data.frame* can be identified via

```
R> rownames(clouds)[clouds$rainfall %in% c(bxpseeding$out,
+      bxpecho$out)]

[1] "1"  "15"
```

where `bxpseeding` and `bxpecho` are variables created by `boxplot` in Figure 5.1. For the time being we shall not remove these observations but bear in mind during the modelling process that they may cause problems.

### *5.3.1 Fitting a Linear Model*

In this example it is sensible to assume that the effect that some of the other explanatory variables is modified by seeding and therefore consider a model that allows interaction terms for `seeding` with each of the covariates except `time`. This model can be described by the *formula*

```
R> clouds_formula <- rainfall ~ seeding * (sne + cloudcover +
+      prewetness + echomotion) + time
```

and the design matrix $\mathbf{X}^\star$ can be computed via

```
R> Xstar <- model.matrix(clouds_formula, data = clouds)
```

By default, treatment contrasts have been applied to the dummy codings of the factors `seeding` and `echomotion` as can be seen from the inspection of the `contrasts` attribute of the model matrix

```
R> attr(Xstar, "contrasts")

$seeding
[1] "contr.treatment"

$echomotion
[1] "contr.treatment"
```

The default contrasts can be changed via the `contrasts.arg` argument to `model.matrix` or the `contrasts` argument to the fitting function, for example `lm` or `aov` as shown in

Chapter 4.

However, such internals are hidden and performed by high-level model fitting functions such as `lm` which will be used to fit the linear model defined by the *formula* `clouds_formula`:

```
R> clouds_lm <- lm(clouds_formula, data = clouds)
R> class(clouds_lm)

[1] "lm"
```

```
R> data("clouds", package = "HSAUR")
R> layout(matrix(1:2, nrow = 2))
R> bxpseeding <- boxplot(rainfall ~ seeding, data = clouds,
+     ylab = "Rainfall", xlab = "Seeding")
R> bxpecho <- boxplot(rainfall ~ echomotion, data = clouds,
+     ylab = "Rainfall", xlab = "Echo Motion")
```

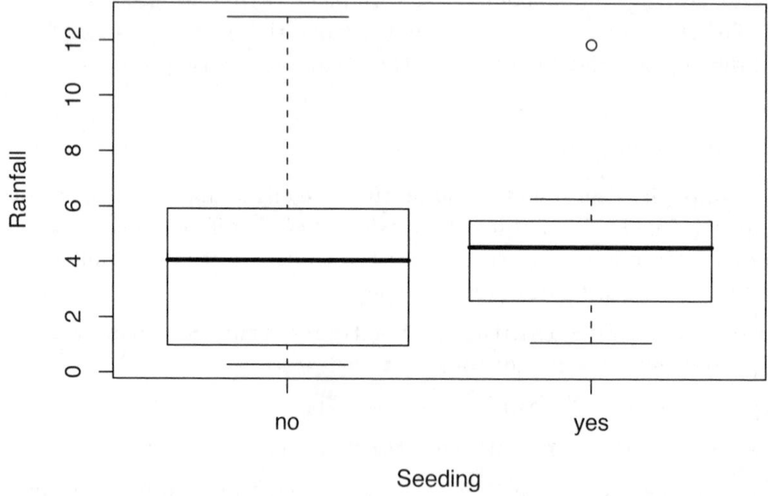


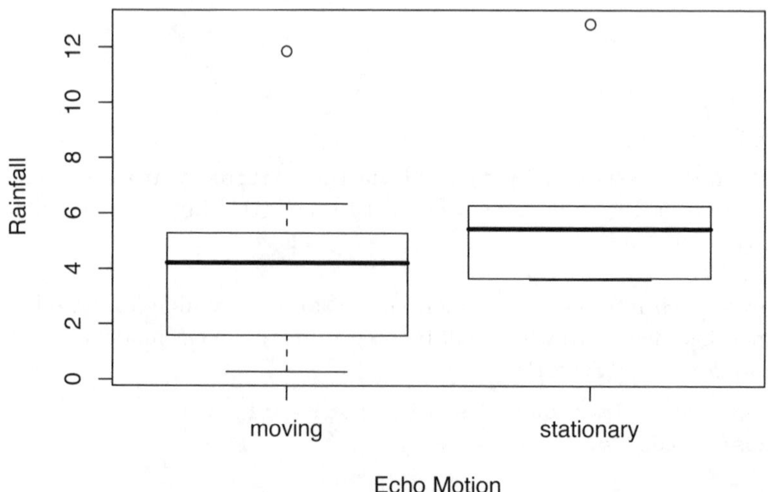


**Figure 5.1** Boxplots of `rainfall`.

```
R> layout(matrix(1:4, nrow = 2))
R> plot(rainfall ~ time, data = clouds)
R> plot(rainfall ~ cloudcover, data = clouds)
R> plot(rainfall ~ sne, data = clouds, xlab = "S-Ne criterion")
R> plot(rainfall ~ prewetness, data = clouds)
```

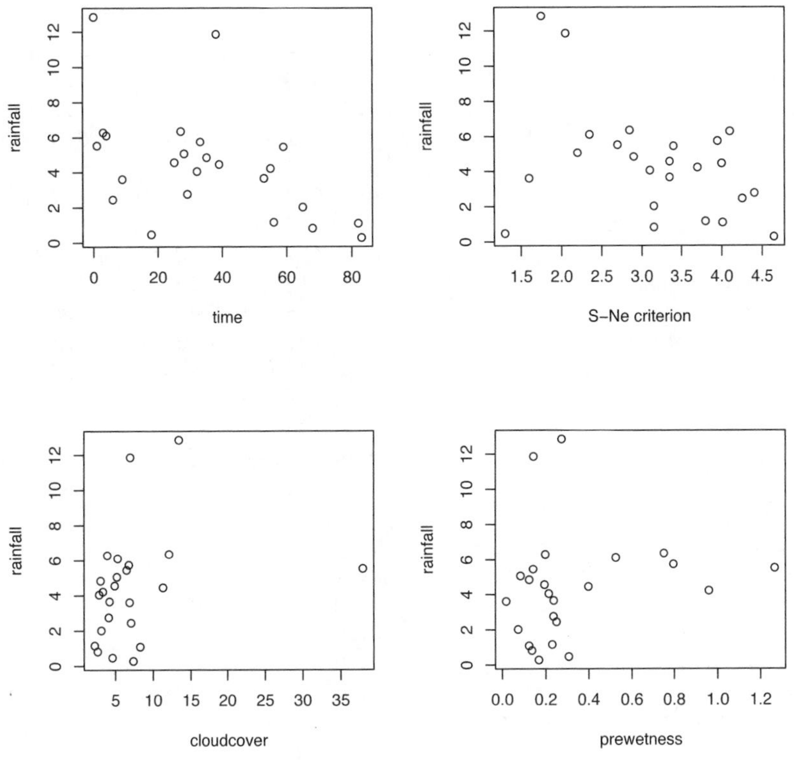


**Figure 5.2** Scatterplots of `rainfall` against the continuous covariates.

The results of the model fitting is an object of class *lm* for which a `summary` method showing the conventional regression analysis output is available. The output in Figure 5.3 shows the estimates $\hat{\beta}^\star$ with corresponding standard errors and $t$-statistics as well as the $F$-statistic with associated $p$-value.

Many methods are available for extracting components of the fitted model. The estimates $\hat{\beta}^\star$ can be assessed via

```
R> betastar <- coef(clouds_lm)
R> betastar
```

```
R> summary(clouds_lm)

Call:
lm(formula = clouds_formula, data = clouds)

Residuals:
    Min      1Q  Median      3Q     Max
-2.5259 -1.1486 -0.2704  1.0401  4.3913

Coefficients:
                                Estimate Std. Error t value
(Intercept)                     -0.34624    2.78773  -0.124
seedingyes                      15.68293    4.44627   3.527
sne                              0.41981    0.84453   0.497
cloudcover                       0.38786    0.21786   1.780
prewetness                       4.10834    3.60101   1.141
echomotionstationary             3.15281    1.93253   1.631
time                            -0.04497    0.02505  -1.795
seedingyes:sne                  -3.19719    1.26707  -2.523
seedingyes:cloudcover           -0.48625    0.24106  -2.017
seedingyes:prewetness           -2.55707    4.48090  -0.571
seedingyes:echomotionstationary -0.56222    2.64430  -0.213
                                Pr(>|t|)
(Intercept)                      0.90306
seedingyes                       0.00372 **
sne                              0.62742
cloudcover                       0.09839 .
prewetness                       0.27450
echomotionstationary             0.12677
time                             0.09590 .
seedingyes:sne                   0.02545 *
seedingyes:cloudcover            0.06482 .
seedingyes:prewetness            0.57796
seedingyes:echomotionstationary  0.83492
---
Signif. codes:  0 '***' 0.001 '**' 0.01 '*' 0.05 '.' 0.1 ' ' 1

Residual standard error: 2.205 on 13 degrees of freedom
Multiple R-Squared: 0.7158,      Adjusted R-squared: 0.4972
F-statistic: 3.274 on 10 and 13 DF,  p-value: 0.02431
```

**Figure 5.3** R output of the linear model fit for the `clouds` data.

```
 (Intercept)
-0.34624093
  seedingyes
15.68293481
         sne
```

```
                  0.41981393
                  cloudcover
                  0.38786207
                  prewetness
                  4.10834188
        echomotionstationary
                  3.15281358
                        time
                 -0.04497427
              seedingyes:sne
                 -3.19719006
       seedingyes:cloudcover
                 -0.48625492
       seedingyes:prewetness
                 -2.55706696
seedingyes:echomotionstationary
                 -0.56221845
```

and the corresponding covariance matrix $\mathsf{Cov}(\hat{\beta}^{\star})$ is available from the `vcov` method

```
R> Vbetastar <- vcov(clouds_lm)
```

where the square roots of the diagonal elements are the standard errors as shown in Figure 5.3

```
R> sqrt(diag(Vbetastar))
```

```
                 (Intercept)
                  2.78773403
                  seedingyes
                  4.44626606
                         sne
                  0.84452994
                  cloudcover
                  0.21785501
                  prewetness
                  3.60100694
        echomotionstationary
                  1.93252592
                        time
                  0.02505286
              seedingyes:sne
                  1.26707204
       seedingyes:cloudcover
                  0.24106012
       seedingyes:prewetness
                  4.48089584
seedingyes:echomotionstationary
                  2.64429975
```

The results of the linear model fit, as shown in Figure 5.3, suggest the interaction of seeding with S-Ne significantly affects rainfall. A suitable graph

```
R> psymb <- as.numeric(clouds$seeding)
R> plot(rainfall ~ sne, data = clouds, pch = psymb,
+      xlab = "S-Ne criterion")
R> abline(lm(rainfall ~ sne, data = clouds,
+      subset = seeding == "no"))
R> abline(lm(rainfall ~ sne, data = clouds,
+      subset = seeding == "yes"), lty = 2)
R> legend("topright", legend = c("No seeding", "Seeding"),
+      pch = 1:2, lty = 1:2, bty = "n")
```

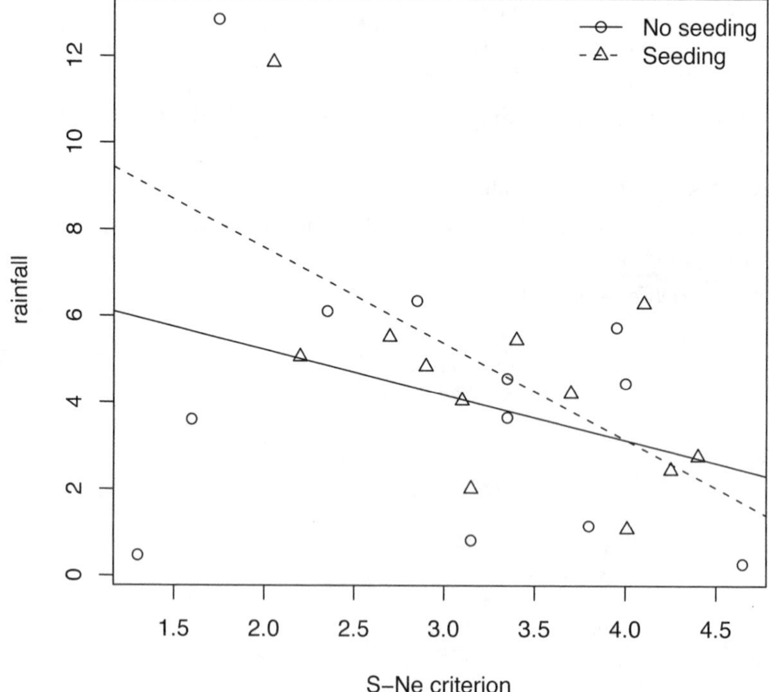


**Figure 5.4** Regression relationship between S-Ne criterion and rainfall with and without seeding.

will help in the interpretation of this result. We can plot the relationship between rainfall and S-Ne for seeding and non-seeding days using the R code shown with Figure 5.4.

The plot suggests that for smaller S-Ne values, seeding produces greater rainfall than no seeding, whereas for larger values of S-Ne it tends to produce less. The cross-over occurs at an S-Ne value of approximately four which suggests that seeding is best carried out when S-Ne is less that four. But the number of observations is small and we should perhaps now consider the influence of any outlying observations on these results.

### *5.3.2 Regression Diagnostics*

The possible influence of outliers and the checking of assumptions made in fitting the multiple regression model, i.e., constant variance and normality of error terms, can both be undertaken using a variety of diagnostic tools, of which the simplest and most well known are the estimated residuals, i.e., the differences between the observed values of the response and the fitted values of the response. So, after estimation, the next stage in the analysis should be an examination of such residuals from fitting the chosen model to check on the normality and constant variance assumptions and to identify outliers. The most useful plots of these residuals are:

- A plot of residuals against each explanatory variable in the model. The presence of a non-linear relationship, for example, may suggest that a higher-order term, in the explanatory variable should be considered.
- A plot of residuals against fitted values. If the variance of the residuals appears to increase with predicted value, a transformation of the response variable may be in order.
- A normal probability plot of the residuals. After all the systematic variation has been removed from the data, the residuals should look like a sample from a standard normal distribution. A plot of the ordered residuals against the expected order statistics from a normal distribution provides a graphical check of this assumption.

In order to investigate the quality of the model fit, we need access to the residuals and the fitted values. The residuals can be found by the `residuals` method and the fitted values of the response from the `fitted` (or `predict`) method

```
R> clouds_resid <- residuals(clouds_lm)
R> clouds_fitted <- fitted(clouds_lm)
```

Now the residuals and the fitted values can be used to construct diagnostic plots; for example the residual plot in Figure 5.5 where each observation is labelled by its number. Observations 1 and 15 give rather large residual values and the data should perhaps be reanalysed after these two observations are removed. The normal probability plot of the residuals shown in Figure 5.6 shows a reasonable agreement between theoretical and sample quantiles, however, observations 1 and 15 are extreme again.

```
R> plot(clouds_fitted, clouds_resid, xlab = "Fitted values",
+      ylab = "Residuals", ylim = max(abs(clouds_resid)) *
+          c(-1, 1), type = "n")
R> abline(h = 0, lty = 2)
R> text(clouds_fitted, clouds_resid, labels = rownames(clouds))
```

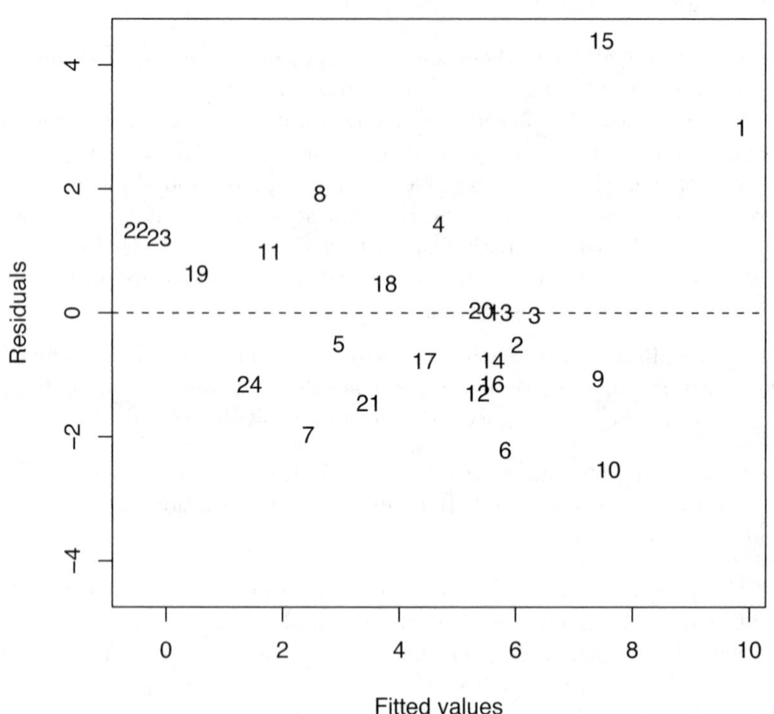


**Figure 5.5** Plot of residuals against fitted values for `clouds` seeding data.

A further diagnostic that is often very useful is an index plot of the Cook's distances for each observation. This statistic is defined as

$$D_k = \frac{1}{(q+1)\hat{\sigma}^2} \sum_{i=1}^{n} (\hat{y}_{i(k)} - y_i)^2$$

where $\hat{y}_{i(k)}$ is the fitted value of the $i$th observation when the $k$th observation is omitted from the model. The values of $D_k$ assess the impact of the $k$th observation on the estimated regression coefficients. Values of $D_k$ greater than

```
R> qqnorm(clouds_resid, ylab = "Residuals")
R> qqline(clouds_resid)
```

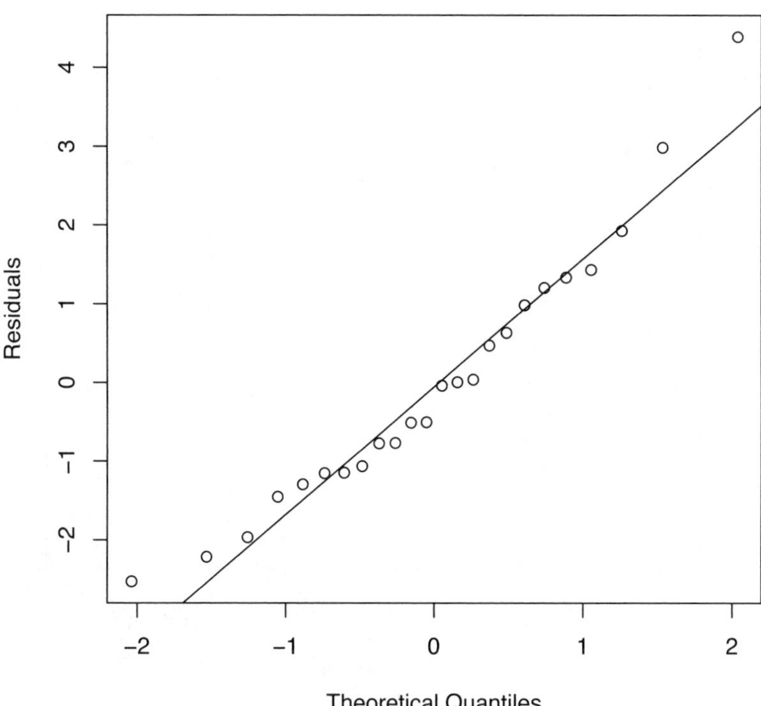


**Figure 5.6** Normal probability plot of residuals from cloud seeding model `clouds_lm`.

one are suggestive that the corresponding observation has undue influence on the estimated regression coefficients (see Cook and Weisberg, 1982).

An index plot of the Cook's distances for each observation (and many other plots including those constructed above from using the basic functions) can be found from applying the `plot` method to the object that results from the application of the `lm` function. Figure 5.7 suggests that observations 2 and 18 have undue influence on the estimated regression coefficients, but the two outliers identified previously do not. Again it may be useful to look at the results after these two observations have been removed (see Exercise 5.2).

```
R> plot(clouds_lm)
```

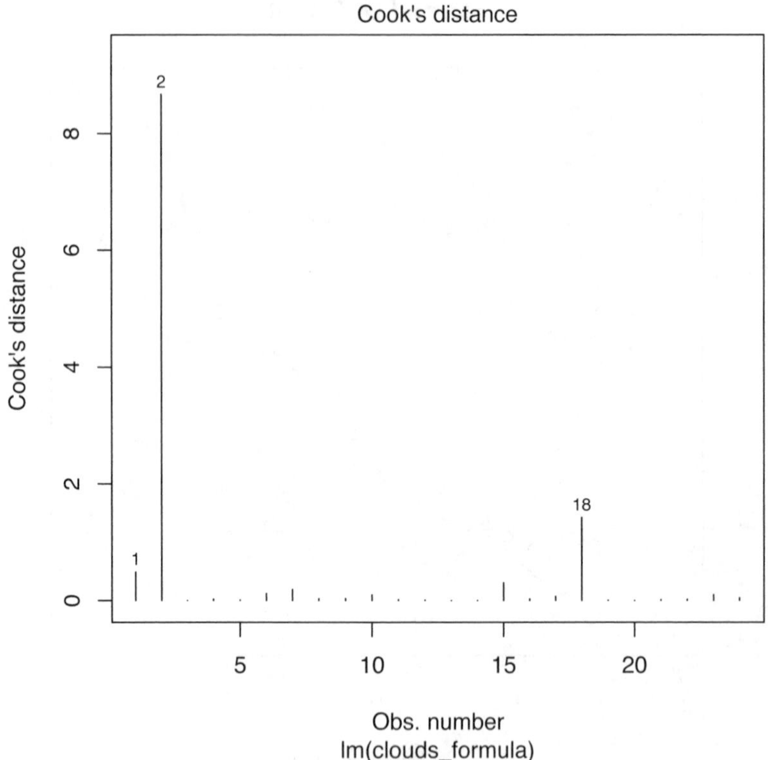


**Figure 5.7** Index plot of Cook's distances for cloud seeding data.

## 5.4 Summary

Multiple regression is used to assess the relationship between a set of explanatory variables and a response variable. The response variable is assumed to be normally distributed with a mean that is a linear function of the explanatory variables and a variance that is independent of the explanatory variables. An important part of any regression analysis involves the graphical examination of residuals and other diagnostic statistics to help identify departures from assumptions.

## Exercises

Ex. 5.1 The simple residuals calculated as the difference between an observed and predicted value have a distribution that is scale dependent since the variance of each is a function of both $\sigma^2$ and the diagonal elements of the *hat matrix* $\mathbf{H}$ given by

$$\mathbf{H} = \mathbf{X}(\mathbf{X}^\top \mathbf{X})^{-1}\mathbf{X}^\top.$$

Consequently it is often more useful to work with the standardised version of the residuals that does not depend on either of these quantities. These standardised residuals are calculated as

$$r_i = \frac{y_i - \hat{y}}{\hat{\sigma}\sqrt{1 - h_{ii}}}$$

where $\hat{\sigma}^2$ is the estimator of $\sigma^2$ and $h_{ii}$ is the $i$th diagonal element of $\mathbf{H}$. Write an R function to calculate these residuals and use it to obtain some diagnostic plots similar to those mentioned in the text. (The elements of the hat matrix can be obtained from the `lm.influence` function.)

Ex. 5.2 Investigate refitting the cloud seeding data after removing any observations which may give cause for concern.

Ex. 5.3 Show how the analysis of variance table for the data in Table 4.1 of the previous chapter can be constructed from the results of applying an appropriate multiple linear regression to the data.

Ex. 5.4 Investigate the use of the `leaps` function from package *leaps* (Lumley and Miller, 2005) for the selecting the 'best' set of variables predicting rainfall in the cloud seeding data.

CHAPTER 6

# Logistic Regression and Generalised Linear Models: Blood Screening, Women's Role in Society, and Colonic Polyps

## 6.1 Introduction

The erythrocyte sedimentation rate (ESR) is the rate at which red blood cells (erythrocytes) settle out of suspension in blood plasma, when measured under standard conditions. If the ESR increases when the level of certain proteins in the blood plasma rise in association with conditions such as rheumatic diseases, chronic infections and malignant diseases, its determination might be useful in screening blood samples taken from people suspected of suffering from one of the conditions mentioned. The absolute value of the ESR is not of great importance, rather it is whether it is less than 20mm/hr since lower values indicate a 'healthy' individual. To ensure that the ESR is a useful diagnostic tool, Collett and Jemain (1985) collected the data shown in Table 6.1.

The question of interest is whether there is any association between the probability of an ESR reading greater than 20mm/hr and the levels of the two plasma proteins. If there is not then the determination of ESR would not be useful for diagnostic purposes.

**Table 6.1**: `plasma` data. Blood plasma data.

| `fibrinogen` | `globulin` | ESR |
|---:|---:|---:|
| 2.52 | 38 | ESR $<$ 20 |
| 2.56 | 31 | ESR $<$ 20 |
| 2.19 | 33 | ESR $<$ 20 |
| 2.18 | 31 | ESR $<$ 20 |
| 3.41 | 37 | ESR $<$ 20 |
| 2.46 | 36 | ESR $<$ 20 |
| 3.22 | 38 | ESR $<$ 20 |
| 2.21 | 37 | ESR $<$ 20 |
| 3.15 | 39 | ESR $<$ 20 |
| 2.60 | 41 | ESR $<$ 20 |
| 2.29 | 36 | ESR $<$ 20 |
| 2.35 | 29 | ESR $<$ 20 |
| 3.15 | 36 | ESR $<$ 20 |

**Table 6.1**: `plasma` data (continued).

| `fibrinogen` | `globulin` | `ESR` |
|---:|---:|---:|
| 2.68 | 34 | ESR < 20 |
| 2.60 | 38 | ESR < 20 |
| 2.23 | 37 | ESR < 20 |
| 2.88 | 30 | ESR < 20 |
| 2.65 | 46 | ESR < 20 |
| 2.28 | 36 | ESR < 20 |
| 2.67 | 39 | ESR < 20 |
| 2.29 | 31 | ESR < 20 |
| 2.15 | 31 | ESR < 20 |
| 2.54 | 28 | ESR < 20 |
| 3.34 | 30 | ESR < 20 |
| 2.99 | 36 | ESR < 20 |
| 3.32 | 35 | ESR < 20 |
| 5.06 | 37 | ESR > 20 |
| 3.34 | 32 | ESR > 20 |
| 2.38 | 37 | ESR > 20 |
| 3.53 | 46 | ESR > 20 |
| 2.09 | 44 | ESR > 20 |
| 3.93 | 32 | ESR > 20 |

*Source*: From Collett, D., Jemain, A., *Sains Malay.*, 4, 493–511, 1985. With permission.

In a survey carried out in 1974/1975 each respondent was asked if he or she agreed or disagreed with the statement 'Women should take care of running their homes and leave running the country up to men'. The responses are summarised in Table 6.2 (from Haberman, 1973) and also given in Collett (2003). The questions here are whether the responses of men and women differ and how years of education affects the response.

**Table 6.2**: `womensrole` data. Women's role in society data.

| `education` | `sex` | `agree` | `disagree` |
|---:|---:|---:|---:|
| 0 | Male | 4 | 2 |
| 1 | Male | 2 | 0 |
| 2 | Male | 4 | 0 |
| 3 | Male | 6 | 3 |
| 4 | Male | 5 | 5 |
| 5 | Male | 13 | 7 |
| 6 | Male | 25 | 9 |
| 7 | Male | 27 | 15 |
| 8 | Male | 75 | 49 |

**Table 6.2**: `womensrole` data (continued).

| `education` | `sex` | `agree` | `disagree` |
|---:|---:|---:|---:|
| 9 | Male | 29 | 29 |
| 10 | Male | 32 | 45 |
| 11 | Male | 36 | 59 |
| 12 | Male | 115 | 245 |
| 13 | Male | 31 | 70 |
| 14 | Male | 28 | 79 |
| 15 | Male | 9 | 23 |
| 16 | Male | 15 | 110 |
| 17 | Male | 3 | 29 |
| 18 | Male | 1 | 28 |
| 19 | Male | 2 | 13 |
| 20 | Male | 3 | 20 |
| 0 | Female | 4 | 2 |
| 1 | Female | 1 | 0 |
| 2 | Female | 0 | 0 |
| 3 | Female | 6 | 1 |
| 4 | Female | 10 | 0 |
| 5 | Female | 14 | 7 |
| 6 | Female | 17 | 5 |
| 7 | Female | 26 | 16 |
| 8 | Female | 91 | 36 |
| 9 | Female | 30 | 35 |
| 10 | Female | 55 | 67 |
| 11 | Female | 50 | 62 |
| 12 | Female | 190 | 403 |
| 13 | Female | 17 | 92 |
| 14 | Female | 18 | 81 |
| 15 | Female | 7 | 34 |
| 16 | Female | 13 | 115 |
| 17 | Female | 3 | 28 |
| 18 | Female | 0 | 21 |
| 19 | Female | 1 | 2 |
| 20 | Female | 2 | 4 |

*Source*: From Haberman, S. J., *Biometrics*, 29, 205–220, 1973. With permission.

Giardiello et al. (1993) and Piantadosi (1997) describe the results of a placebo-controlled trial of a non-steroidal anti-inflammatory drug in the treatment of familial andenomatous polyposis (FAP). The trial was halted after a planned interim analysis had suggested compelling evidence in favour of the treatment. The data shown in Table 6.3 give the number of colonic polyps after a 12-month treatment period. The question of interest is whether the number of polyps is related to treatment and/or age of patients.

**Table 6.3**: `polyps` data. Number of polyps for two treatment arms.

| number | treat | age | number | treat | age |
|---|---|---|---|---|---|
| 63 | placebo | 20 | 3 | drug | 23 |
| 2 | drug | 16 | 28 | placebo | 22 |
| 28 | placebo | 18 | 10 | placebo | 30 |
| 17 | drug | 22 | 40 | placebo | 27 |
| 61 | placebo | 13 | 33 | drug | 23 |
| 1 | drug | 23 | 46 | placebo | 22 |
| 7 | placebo | 34 | 50 | placebo | 34 |
| 15 | placebo | 50 | 3 | drug | 23 |
| 44 | placebo | 19 | 1 | drug | 22 |
| 25 | drug | 17 | 4 | drug | 42 |

## 6.2 Logistic Regression and Generalised Linear Models

### *6.2.1 Logistic Regression*

One way of writing the multiple regression model described in the previous chapter is as $y \sim \mathcal{N}(\mu, \sigma^2)$ where $\mu = \beta_0 + \beta_1 x_1 + \ldots + \beta_q x_q$. This makes it clear that this model is suitable for continuous response variables with, conditional on the values of the explanatory variables, a normal distribution with constant variance. So clearly the model would not be suitable for applying to the erythrocyte sedimentation rate in Table 6.1, since the response variable is binary. If we were to model the expected value of this type of response, i.e., the probability of it taking the value one, say $\pi$, directly as a linear function of explanatory variables, it could lead to fitted values of the response probability outside the range $[0, 1]$, which would clearly not be sensible. And if we write the value of the binary response as $y = \pi(x_1, x_2, \ldots, x_q) + \varepsilon$ it soon becomes clear that the assumption of normality for $\varepsilon$ is also wrong. In fact here $\varepsilon$ may assume only one of two possible values. If $y = 1$, then $\varepsilon = 1 - \pi(x_1, x_2, \ldots, x_q)$ with probability $\pi(x_1, x_2, \ldots, x_q)$ and if $y = 0$ then $\varepsilon = \pi(x_1, x_2, \ldots, x_q)$ with probability $1 - \pi(x_1, x_2, \ldots, x_q)$. So $\varepsilon$ has a distribution with mean zero and variance equal to $\pi(x_1, x_2, \ldots, x_q)(1 - \pi(x_1, x_2, \ldots, x_q))$, i.e., the conditional distribution of our binary response variable follows a binomial distribution with probability given by the conditional mean, $\pi(x_1, x_2, \ldots, x_q)$.

So instead of modelling the expected value of the response directly as a linear function of explanatory variables, a suitable transformation is modelled. In this case the most suitable transformation is the *logistic* or *logit* function of $\pi$ leading to the model

$$\text{logit}(\pi) = \log\left(\frac{\pi}{1-\pi}\right) = \beta_0 + \beta_1 x_1 + \ldots + \beta_q x_q. \tag{6.1}$$

The logit of a probability is simply the log of the odds of the response taking the value one. Equation (6.1) can be rewritten as

$$\pi(x_1, x_2, \ldots, x_q) = \frac{\exp(\beta_0 + \beta_1 x_1 + \ldots + \beta_q x_q)}{1 + \exp(\beta_0 + \beta_1 x_1 + \ldots + \beta_q x_q)}. \tag{6.2}$$

The logit function can take any real value, but the associated probability always lies in the required $[0, 1]$ interval. In a logistic regression model, the parameter $\beta_j$ associated with explanatory variable $x_j$ is such that $\exp(\beta_j)$ is the odds that the response variable takes the value one when $x_j$ increases by one, conditional on the other explanatory variables remaining constant. The parameters of the logistic regression model (the vector of regression coefficients $\beta$) are estimated by maximum likelihood; details are given in Collett (2003).

### *6.2.2 The Generalised Linear Model*

The analysis of variance models considered in Chapter 4 and the multiple regression model described in Chapter 5 are, essentially, completely equivalent. Both involve a linear combination of a set of explanatory variables (dummy variables in the case of analysis of variance) as a model for the observed response variable. And both include residual terms assumed to have a normal distribution. The equivalence of analysis of variance and multiple regression is spelt out in more detail in Everitt (2001).

The logistic regression model described in this chapter also has similarities to the analysis of variance and multiple regression models. Again a linear combination of explanatory variables is involved, although here the expected value of the binary response is not modelled directly but via a logistic transformation. In fact all three techniques can be unified in the *generalised linear model* (GLM), first introduced in a landmark paper by Nelder and Wedderburn (1972). The GLM enables a wide range of seemingly disparate problems of statistical modelling and inference to be set in an elegant unifying framework of great power and flexibility. A comprehensive technical account of the model is given in McCullagh and Nelder (1989). Here we describe GLMs only briefly. Essentially GLMs consist of three main features;

1. An *error distribution* giving the distribution of the response around its mean. For analysis of variance and multiple regression this will be the normal; for logistic regression it is the binomial. Each of these (and others used in other situations to be described later) come from the same, *exponential family* of probability distributions, and it is this family that is used in generalised linear modelling (see Everitt and Pickles, 2000).

2. A *link function*, $g$, that shows how the linear function of the explanatory

variables is related to the expected value of the response

$$g(\mu) = \beta_0 + \beta_1 x_1 + \ldots + \beta_q x_q.$$

For analysis of variance and multiple regression the link function is simply the identity function; in logistic regression it is the logit function.

3. The *variance function* that captures how the variance of the response variable depends on the mean. We will return to this aspect of GLMs later in the chapter.

Estimation of the parameters in a GLM is usually achieved through a maximum likelihood approach – see McCullagh and Nelder (1989) for details. Having estimated a GLM for a data set, the question of the quality of its fit arises. Clearly the investigator needs to be satisfied that the chosen model describes the data adequately, before drawing conclusions about the parameter estimates themselves. In practice, most interest will lie in comparing the fit of competing models, particularly in the context of selecting subsets of explanatory variables that describe the data in a parsimonious manner. In GLMs a measure of fit is provided by a quantity known as the deviance which measures how closely the model-based fitted values of the response approximate the observed value. Comparing the deviance values for two models gives a likelihood ratio test of the two models that can be compared by using a statistic having a $\chi^2$-distribution with degrees of freedom equal to the difference in the number of parameters estimated under each model. More details are given in Cook (1998).

## 6.3 Analysis Using R

### *6.3.1 ESR and Plasma Proteins*

We begin by looking at the ESR data from Table 6.1. As always it is good practice to begin with some simple graphical examination of the data before undertaking any formal modelling. Here we will look at conditional density plots of the response variable given the two explanatory variables describing how the conditional distribution of the categorical variable ESR changes over the numerical variables fibrinogen and gamma globulin. The required R code to construct these plots is shown with Figure 6.1. It appears that higher levels of each protein are associated with ESR values above 20 mm/hr.

We can now fit a logistic regression model to the data using the `glm` function. We start with a model that includes only a single explanatory variable, `fibrinogen`. The code to fit the model is

```
R> plasma_glm_1 <- glm(ESR ~ fibrinogen, data = plasma,
+      family = binomial())
```

The formula implicitly defines a parameter for the global mean (the intercept term) as discussed in Chapters 4 and 5. The distribution of the response is defined by the `family` argument, a binomial distribution in our case. (The

```
R> data("plasma", package = "HSAUR")
R> layout(matrix(1:2, ncol = 2))
R> cdplot(ESR ~ fibrinogen, data = plasma)
R> cdplot(ESR ~ globulin, data = plasma)
```

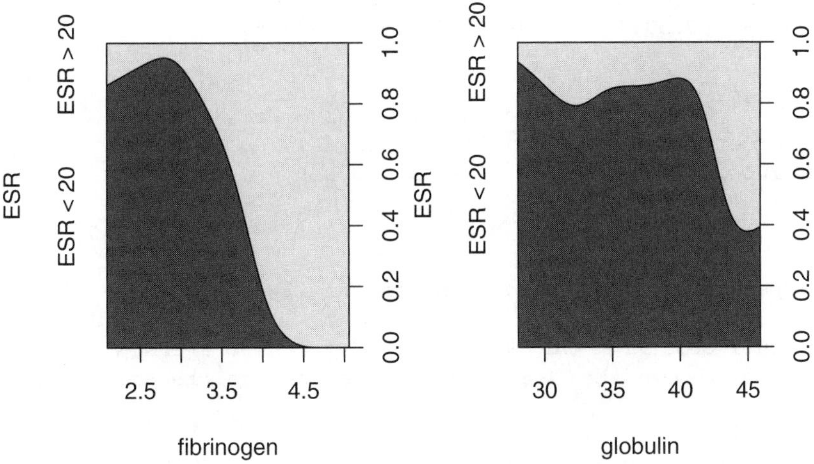


**Figure 6.1** Conditional density plots of the erythrocyte sedimentation rate (ESR) given fibrinogen and globulin.

default link function when the binomial family is requested is the logistic function.)

A description of the fitted model can be obtained from the summary method applied to the fitted model. The output is shown in Figure 6.2.

From the results in Figure 6.2 we see that the regression coefficient for fibrinogen is significant at the 5% level. An increase of one unit in this variable increases the log-odds in favour of an ESR value greater than 20 by an estimated 1.83 with 95% confidence interval

```
R> confint(plasma_glm_1, parm = "fibrinogen")
    2.5 %    97.5 %
0.3389465 3.9988602
```

These values are more helpful if converted to the corresponding values for the odds themselves by exponentiating the estimate

```
R> exp(coef(plasma_glm_1)["fibrinogen"])
fibrinogen
  6.215715
```

and the confidence interval

```
R> summary(plasma_glm_1)

Call:
glm(formula = ESR ~ fibrinogen, family = binomial(), data =
    plasma)

Deviance Residuals:
    Min       1Q   Median       3Q      Max
-0.9298  -0.5399  -0.4382  -0.3356   2.4794

Coefficients:
            Estimate Std. Error z value Pr(>|z|)
(Intercept)  -6.8451     2.7703  -2.471   0.0135 *
fibrinogen    1.8271     0.9009   2.028   0.0425 *
---
Signif. codes:  0 '***' 0.001 '**' 0.01 '*' 0.05 '.' 0.1 ' ' 1

(Dispersion parameter for binomial family taken to be 1)

    Null deviance: 30.885  on 31  degrees of freedom
Residual deviance: 24.840  on 30  degrees of freedom
AIC: 28.840

Number of Fisher Scoring iterations: 5
```

**Figure 6.2** R output of the summary method for the logistic regression model fitted to the plasma data.

```
R> exp(confint(plasma_glm_1, parm = "fibrinogen"))

    2.5 %     97.5 %
 1.403468 54.535954
```

The confidence interval is very wide because there are few observations overall and very few where the ESR value is greater than 20. Nevertheless it seems likely that increased values of fibrinogen lead to a greater probability of an ESR value greater than 20.

We can now fit a logistic regression model that includes both explanatory variables using the code

```
R> plasma_glm_2 <- glm(ESR ~ fibrinogen + globulin,
+      data = plasma, family = binomial())
```

and the output of the summary method is shown in Figure 6.3.

The coefficient for gamma globulin is not significantly different from zero. Subtracting the residual deviance of the second model from the corresponding value for the first model we get a value of 1.87. Tested using a $\chi^2$-distribution with a single degree of freedom this is not significant at the 5% level and so we conclude that gamma globulin is not associated with ESR level. In R, the

```
R> summary(plasma_glm_2)

Call:
glm(formula = ESR ~ fibrinogen + globulin, family = binomial(),
    data = plasma)

Deviance Residuals:
    Min       1Q   Median       3Q      Max
-0.9683  -0.6122  -0.3458  -0.2116   2.2636

Coefficients:
            Estimate Std. Error z value Pr(>|z|)
(Intercept) -12.7921     5.7963  -2.207   0.0273 *
fibrinogen    1.9104     0.9710   1.967   0.0491 *
globulin      0.1558     0.1195   1.303   0.1925
---
Signif. codes:  0 '***' 0.001 '**' 0.01 '*' 0.05 '.' 0.1 ' ' 1

(Dispersion parameter for binomial family taken to be 1)

    Null deviance: 30.885  on 31  degrees of freedom
Residual deviance: 22.971  on 29  degrees of freedom
AIC: 28.971

Number of Fisher Scoring iterations: 5
```

**Figure 6.3** R output of the summary method for the logistic regression model fitted to the plasma data.

task of comparing the two nested models can be performed using the anova function

```
R> anova(plasma_glm_1, plasma_glm_2, test = "Chisq")
Analysis of Deviance Table

Model 1: ESR ~ fibrinogen
Model 2: ESR ~ fibrinogen + globulin
  Resid. Df Resid. Dev Df Deviance P(>|Chi|)
1        30    24.8404
2        29    22.9711  1   1.8692    0.1716
```

Nevertheless we shall use the predicted values from the second model and plot them against the values of *both* explanatory variables using a *bubble plot* to illustrate the use of the symbols function. The estimated conditional probability of a ESR value larger 20 for all observations can be computed, following formula (6.2), by

```
R> prob <- predict(plasma_glm_1, type = "response")
```

and now we can assign a larger circle to observations with larger probability

```
R> plot(globulin ~ fibrinogen, data = plasma, xlim = c(2,
+     6), ylim = c(25, 50), pch = ".")
R> symbols(plasma$fibrinogen, plasma$globulin, circles = prob,
+     add = TRUE)
```

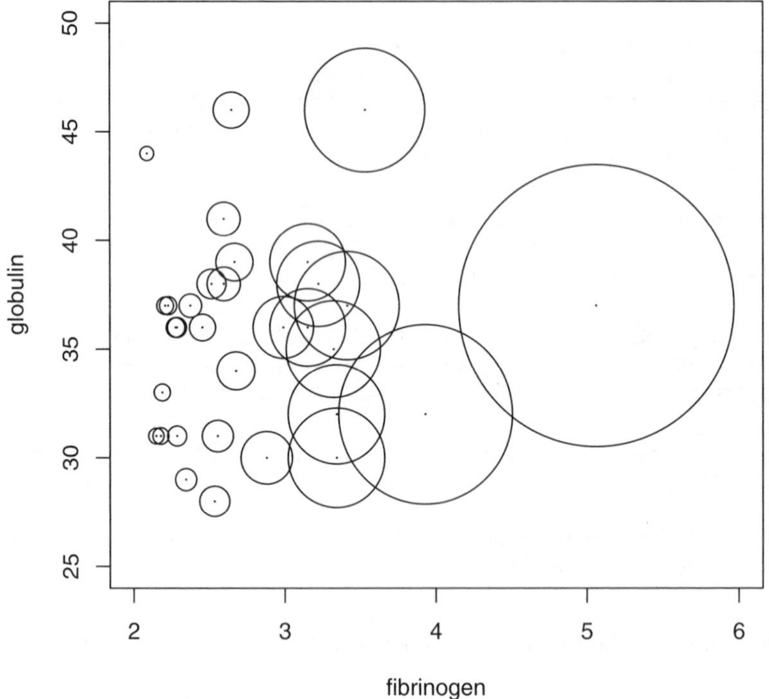


**Figure 6.4** Bubble plot of fitted values for a logistic regression model fitted to the ESR data.

as shown in Figure 6.4. The plot clearly shows the increasing probability of an ESR value above 20 (larger circles) as the values of fibrinogen, and to a lesser extent, gamma globulin, increase.

### *6.3.2 Women's Role in Society*

Originally the data in Table 6.2 would have been in a completely equivalent form to the data in Table 6.1 data, but here the individual observations have been grouped into counts of numbers of agreements and disagreements for the two explanatory variables, `sex` and `education`. To fit a logistic regression model to such grouped data using the `glm` function we need to specify the number of agreements and disagreements as a two-column matrix on the left hand side of the model formula. We first fit a model that includes the two explanatory variables using the code

```
R> data("womensrole", package = "HSAUR")
R> womensrole_glm_1 <- glm(cbind(agree, disagree) ~
+      sex + education, data = womensrole, family = binomial())
```

```
R> summary(womensrole_glm_1)

Call:
glm(formula = cbind(agree, disagree) ~ sex + education,
    family = binomial(), data = womensrole)

Deviance Residuals:
     Min        1Q    Median        3Q       Max
-2.72544  -0.86302  -0.06525   0.84340   3.13315

Coefficients:
            Estimate Std. Error z value Pr(>|z|)
(Intercept)  2.50937    0.18389  13.646   <2e-16 ***
sexFemale   -0.01145    0.08415  -0.136    0.892
education   -0.27062    0.01541 -17.560   <2e-16 ***
---
Signif. codes:  0 '***' 0.001 '**' 0.01 '*' 0.05 '.' 0.1 ' ' 1

(Dispersion parameter for binomial family taken to be 1)

    Null deviance: 451.722  on 40  degrees of freedom
Residual deviance:  64.007  on 38  degrees of freedom
AIC: 208.07

Number of Fisher Scoring iterations: 4
```

**Figure 6.5** R output of the `summary` method for the logistic regression model fitted to the `womensrole` data.

From the `summary` output in Figure 6.5 it appears that education has a highly significant part to play in predicting whether a respondent will agree with the statement read to them, but the respondent's sex is apparently unimportant. As years of education increase the probability of agreeing with the statement declines. We now are going to construct a plot comparing the observed proportions of agreeing with those fitted by our fitted model. Because we will reuse this plot for another fitted object later on, we define a function which plots years of education against some fitted probabilities, e.g.,

```
R> role.fitted1 <- predict(womensrole_glm_1, type = "response")
```

and labels each observation with the person's sex:

```
R> myplot <- function(role.fitted) {
+     f <- womensrole$sex == "Female"
+     plot(womensrole$education, role.fitted, type = "n",
+         ylab = "Probability of agreeing", xlab = "Education",
+         ylim = c(0, 1))
+     lines(womensrole$education[!f], role.fitted[!f],
+         lty = 1)
+     lines(womensrole$education[f], role.fitted[f],
+         lty = 2)
+     lgtxt <- c("Fitted (Males)", "Fitted (Females)")
+     legend("topright", lgtxt, lty = 1:2, bty = "n")
+     y <- womensrole$agree/
+         (womensrole$agree + womensrole$disagree)
+     text(womensrole$education, y, ifelse(f, "\\VE",
+         "\\MA"), family = "HersheySerif", cex = 1.25)
+ }
```

The two curves for males and females in Figure 6.6 are almost the same reflecting the non-significant value of the regression coefficient for sex in womensrole_glm_1. But the observed values plotted on Figure 6.6 suggest that there might be an interaction of education and sex, a possibility that can be investigated by applying a further logistic regression model using

```
R> womensrole_glm_2 <- glm(cbind(agree, disagree) ~
+     sex * education, data = womensrole, family = binomial())
```

The sex and education interaction term is seen to be highly significant, as can be seen from the summary output in Figure 6.7.

Interpreting this interaction effect is made simpler if we again plot fitted and observed values using the same code as previously after getting fitted values from womensrole_glm_2. The plot is shown in Figure 6.8. We see that for fewer years of education women have a higher probability of agreeing with the statement than men, but when the years of education exceed about ten then this situation reverses.

A range of residuals and other diagnostics is available for use in association with logistic regression to check whether particular components of the model are adequate. A comprehensive account of these is given in Collett (2003); here we shall demonstrate only the use of what is known as the *deviance residual*. This is the signed square root of the contribution of the $i$th observation to the overall deviance. Explicitly it is given by

$$d_i = \text{sign}(y_i - \hat{y}_i)\left(2y_i \log\left(\frac{y_i}{\hat{y}_i}\right) + 2(n_i - y_i)\log\left(\frac{n_i - y_i}{n_i - \hat{y}_i}\right)\right)^{1/2} \qquad (6.3)$$

where sign is the function that makes $d_i$ positive when $y_i \geq \hat{y}_i$ and negative else. In (6.3) $y_i$ is the observed number of ones for the $i$th observation (the number of people who agree for each combination of covariates in our example), and $\hat{y}_i$ is its fitted value from the model. The residual provides information about how well the model fits each particular observation.

```
R> myplot(role.fitted1)
```

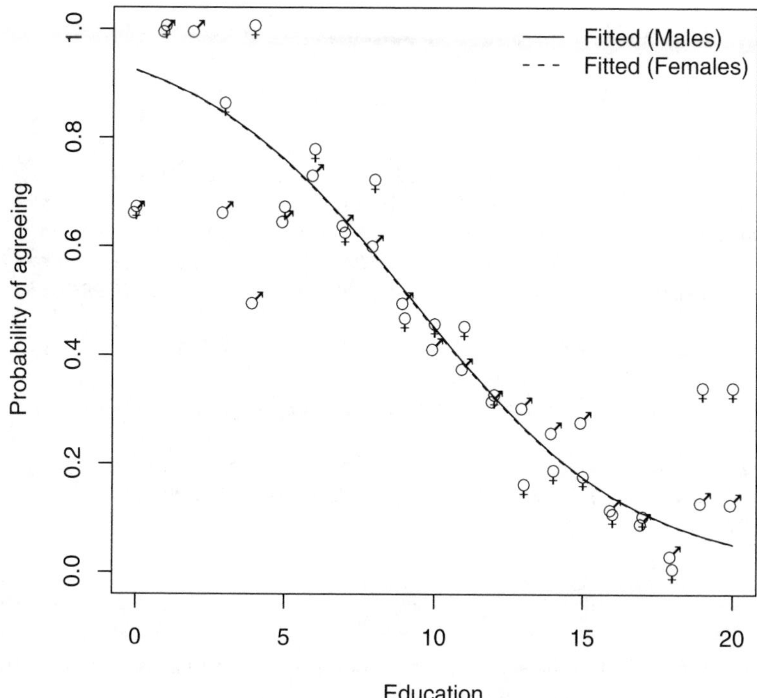


**Figure 6.6** Fitted (from `womensrole_glm_1`) and observed probabilities of agreeing for the `womensrole` data.

We can obtain a plot of deviance residuals plotted against fitted values using the following code above Figure 6.9. The residuals fall into a horizontal band between −2 and 2. This pattern does not suggest a poor fit for any particular observation or subset of observations.

### *6.3.3 Colonic Polyps*

The data on colonic polyps in Table 6.3 involves *count* data. We could try to model this using multiple regression but there are two problems. The first is that a response that is a count can only take positive values, and secondly such a variable is unlikely to have a normal distribution. Instead we will apply a GLM with a log link function, ensuring that fitted values are positive, and

```
R> summary(womensrole_glm_2)

Call:
glm(formula = cbind(agree, disagree) ~ sex * education,
    family = binomial(), data = womensrole)

Deviance Residuals:
     Min        1Q    Median        3Q       Max
-2.39097  -0.88062   0.01532   0.72783   2.45262

Coefficients:
                    Estimate Std. Error z value Pr(>|z|)
(Intercept)          2.09820    0.23550   8.910  < 2e-16 ***
sexFemale            0.90474    0.36007   2.513  0.01198 *
education           -0.23403    0.02019 -11.592  < 2e-16 ***
sexFemale:education -0.08138    0.03109  -2.617  0.00886 **
---
Signif. codes:  0 '***' 0.001 '**' 0.01 '*' 0.05 '.' 0.1 ' ' 1

(Dispersion parameter for binomial family taken to be 1)

    Null deviance: 451.722  on 40  degrees of freedom
Residual deviance:  57.103  on 37  degrees of freedom
AIC: 203.16

Number of Fisher Scoring iterations: 4
```

**Figure 6.7** R output of the `summary` method for the logistic regression model fitted to the `womensrole` data.

a Poisson error distribution, i.e.,

$$\mathsf{P}(y) = \frac{e^{-\lambda}\lambda^y}{y!}.$$

This type of GLM is often known as *Poisson regression*. We can apply the model using

```
R> data("polyps", package = "HSAUR")
R> polyps_glm_1 <- glm(number ~ treat + age, data = polyps,
+      family = poisson())
```

(The default link function when the Poisson family is requested is the log function.)

From Figure 6.10 we see that the regression coefficients for both age and treatment are highly significant. But there is a problem with the model, but before we can deal with it we need a short digression to describe in more detail the third component of GLMs mentioned in the previous section, namely their variance functions, $V(\mu)$.

```
R> role.fitted2 <- predict(womensrole_glm_2, type = "response")
R> myplot(role.fitted2)
```

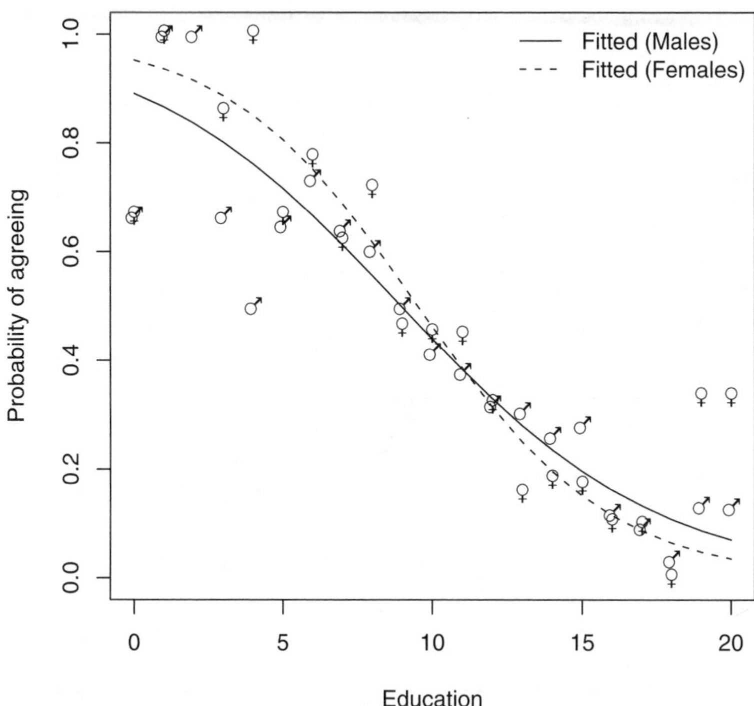


**Figure 6.8** Fitted (from `womensrole_glm_2`) and observed probabilities of agreeing for the `womensrole` data.

The variance function of a GLM captures how the variance of a response variable depends upon its mean. The general form of the relationship is

$$\mathsf{Var}(\text{response}) = \phi V(\mu)$$

where $\phi$ is constant and $V(\mu)$ specifies how the variance depends on the mean. For the error distributions considered previously this general form becomes:

**Normal:** $V(\mu) = 1, \phi = \sigma^2$; here the variance does not depend on the mean.

**Binomial:** $V(\mu) = \mu(1 - \mu), \phi = 1$.

**Poisson:** $V(\mu) = \mu, \phi = 1$.

```
R> res <- residuals(womensrole_glm_2, type = "deviance")
R> plot(predict(womensrole_glm_2), res, xlab = "Fitted values",
+      ylab = "Residuals", ylim = max(abs(res)) * c(-1,
+          1))
R> abline(h = 0, lty = 2)
```

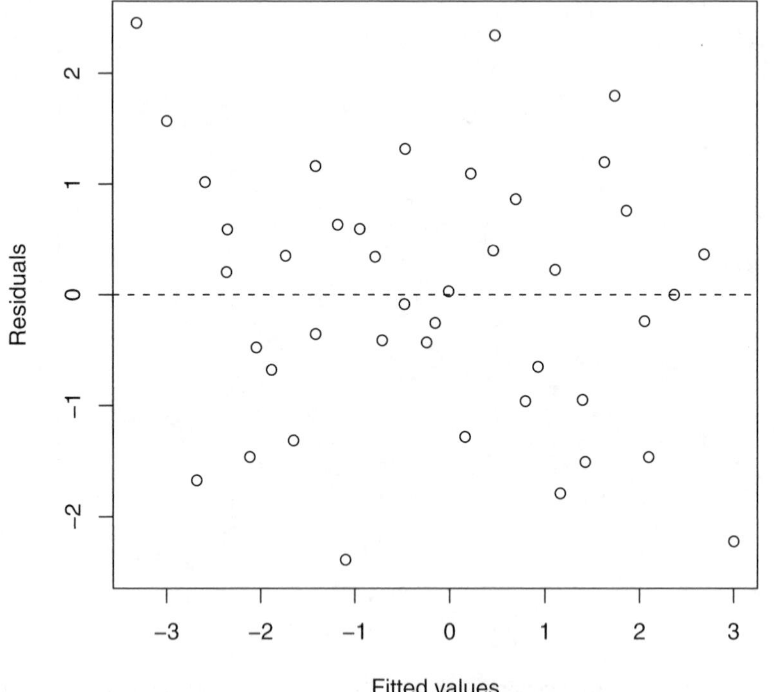


**Figure 6.9** Plot of deviance residuals from logistic regression model fitted to the `womensrole` data.

In the case of a Poisson variable we see that the mean and variance are equal, and in the case of a binomial variable where the mean is the probability of the variable taking the value one, $\pi$, the variance is $\pi(1 - \pi)$.

Both the Poisson and binomial distributions have variance functions that are completely determined by the mean. There is no free parameter for the variance since, in applications of the generalised linear model with binomial or Poisson error distributions the dispersion parameter, $\phi$, is defined to be one (see previous results for logistic and Poisson regression). But in some applica-

```
R> summary(polyps_glm_1)

Call:
glm(formula = number ~ treat + age, family = poisson(), data =
    polyps)

Deviance Residuals:
    Min       1Q   Median       3Q      Max
-4.2212  -3.0536  -0.1802   1.4459   5.8301

Coefficients:
             Estimate Std. Error z value Pr(>|z|)
(Intercept)  4.529024   0.146872   30.84  < 2e-16 ***
treatdrug   -1.359083   0.117643  -11.55  < 2e-16 ***
age         -0.038830   0.005955   -6.52 7.02e-11 ***
---
Signif. codes:  0 '***' 0.001 '**' 0.01 '*' 0.05 '.' 0.1 ' ' 1

(Dispersion parameter for poisson family taken to be 1)

    Null deviance: 378.66  on 19  degrees of freedom
Residual deviance: 179.54  on 17  degrees of freedom
AIC: 273.88

Number of Fisher Scoring iterations: 5
```

**Figure 6.10** R output of the summary method for the Poisson regression model fitted to the polyps data.

tions this becomes too restrictive to fully account for the empirical variance in the data; in such cases it is common to describe the phenomenon as *overdispersion*. For example, if the response variable is the proportion of family members who have been ill in the past year, observed in a large number of families, then the individual binary observations that make up the observed proportions are likely to be correlated rather than independent. The non-independence can lead to a variance that is greater (less) than on the assumption of binomial variability. And observed counts often exhibit larger variance than would be expected from the Poisson assumption, a fact noted over 80 years ago by Greenwood and Yule (1920).

When fitting generalised models with binomial or Poisson error distributions, overdispersion can often be spotted by comparing the residual deviance with its degrees of freedom. For a well-fitting model the two quantities should be approximately equal. If the deviance is far greater than the degrees of freedom overdispersion may be indicated. This is the case for the results in Figure 6.10. So what can we do?

We can deal with overdispersion by using a procedure known as *quasi-likelihood*, which allows the estimation of model parameters without fully knowing the error distribution of the response variable. McCullagh and Nelder (1989) give full details of the quasi-likelihood approach. In many respects it simply allows for the estimation of $\phi$ from the data rather than defining it to be unity for the binomial and Poisson distributions. We can apply quasi-likelihood estimation to the colonic polyps data using the following R code

```
R> polyps_glm_2 <- glm(number ~ treat + age, data = polyps,
+      family = quasipoisson())
R> summary(polyps_glm_2)
```

```
Call:
glm(formula = number ~ treat + age, family = quasipoisson(),
    data = polyps)

Deviance Residuals:
    Min       1Q   Median       3Q      Max
-4.2212  -3.0536  -0.1802   1.4459   5.8301

Coefficients:
            Estimate Std. Error t value Pr(>|t|)
(Intercept)  4.52902    0.48105   9.415 3.72e-08 ***
treatdrug   -1.35908    0.38532  -3.527  0.00259 **
age         -0.03883    0.01951  -1.991  0.06283 .
---
Signif. codes:  0 '***' 0.001 '**' 0.01 '*' 0.05 '.' 0.1 ' ' 1

(Dispersion parameter for quasipoisson family taken to be
    10.72761)

    Null deviance: 378.66  on 19  degrees of freedom
Residual deviance: 179.54  on 17  degrees of freedom
AIC: NA

Number of Fisher Scoring iterations: 5
```

The regression coefficients for both explanatory variables remain significant but their estimated standard errors are now much greater than the values given in Figure 6.10. A possible reason for overdispersion in these data is that polyps do not occur independently of one another, but instead may 'cluster' together.

## 6.4 Summary

Generalised linear models provide a very powerful and flexible framework for the application of regression models to a variety of non-normal response variables, for example, logistic regression to binary responses and Poisson regression to count data.

**Exercises**

Ex. 6.1 Construct a perspective plot of the fitted values from a logistic regression model fitted to the `plasma` data in which both fibrinogen and gamma globulin are included as explanatory variables.

Ex. 6.2 Collett (2003) argues that two outliers need to be removed from the `plasma` data. Try to identify those two unusual observations by means of a scatterplot.

Ex. 6.3 The data shown in Table 6.4 arise from 31 male patients who have been treated for superficial bladder cancer (see Seeber, 1998), and give the number of recurrent tumours during a particular time after the removal of the primary tumour, along with the size of the original tumour (whether smaller or larger than 3 cm). Use Poisson regression to estimate the effect of size of tumour on the number of recurrent tumours.

**Table 6.4**: `bladdercancer` data. Number of recurrent tumours for bladder cancer patients.

| time | tumorsize | number | time | tumorsize | number |
|---:|---:|---:|---:|---:|---:|
| 2 | <=3cm | 1 | 13 | <=3cm | 2 |
| 3 | <=3cm | 1 | 15 | <=3cm | 2 |
| 6 | <=3cm | 1 | 18 | <=3cm | 2 |
| 8 | <=3cm | 1 | 23 | <=3cm | 2 |
| 9 | <=3cm | 1 | 20 | <=3cm | 3 |
| 10 | <=3cm | 1 | 24 | <=3cm | 4 |
| 11 | <=3cm | 1 | 1 | >3cm | 1 |
| 13 | <=3cm | 1 | 5 | >3cm | 1 |
| 14 | <=3cm | 1 | 17 | >3cm | 1 |
| 16 | <=3cm | 1 | 18 | >3cm | 1 |
| 21 | <=3cm | 1 | 25 | >3cm | 1 |
| 22 | <=3cm | 1 | 18 | >3cm | 2 |
| 24 | <=3cm | 1 | 25 | >3cm | 2 |
| 26 | <=3cm | 1 | 4 | >3cm | 3 |
| 27 | <=3cm | 1 | 19 | >3cm | 4 |
| 7 | <=3cm | 2 | | | |

*Source*: From Seeber, G. U. H., in *Encyclopedia of Biostatistics*, John Wiley & Sons, Chichester, UK, 1998. With permission.

Ex. 6.4 The data in Table 6.5 show the survival times from diagnosis of patients suffering from leukemia and the values of two explanatory variables, the white blood cell count (`wbc`) and the presence or absence of a morphological characteristic of the white blood cells (`ag`) (the data are available in package *MASS*, Venables and Ripley, 2002). Define a binary outcome variable according to whether or not patients lived for at least 24 weeks after diagnosis and then fit a logistic regression model to the data. It may be advisable to transform the very large white blood counts to avoid regression coefficients very close to 0 (and odds ratios very close to 1). And a model that contains only the two explanatory variables may not be adequate for these data. Construct some graphics useful in the interpretation of the final model you fit.

**Table 6.5**: `leuk` data (package *MASS*). Survival times of patients suffering from leukemia.

| wbc | ag | time | wbc | ag | time |
|---|---|---|---|---|---|
| 2300 | present | 65 | 4400 | absent | 56 |
| 750 | present | 156 | 3000 | absent | 65 |
| 4300 | present | 100 | 4000 | absent | 17 |
| 2600 | present | 134 | 1500 | absent | 7 |
| 6000 | present | 16 | 9000 | absent | 16 |
| 10500 | present | 108 | 5300 | absent | 22 |
| 10000 | present | 121 | 10000 | absent | 3 |
| 17000 | present | 4 | 19000 | absent | 4 |
| 5400 | present | 39 | 27000 | absent | 2 |
| 7000 | present | 143 | 28000 | absent | 3 |
| 9400 | present | 56 | 31000 | absent | 8 |
| 32000 | present | 26 | 26000 | absent | 4 |
| 35000 | present | 22 | 21000 | absent | 3 |
| 100000 | present | 1 | 79000 | absent | 30 |
| 100000 | present | 1 | 100000 | absent | 4 |
| 52000 | present | 5 | 100000 | absent | 43 |
| 100000 | present | 65 | | | |

CHAPTER 7

# Density Estimation: Erupting Geysers and Star Clusters

## 7.1 Introduction

Geysers are natural fountains that shoot up into the air, at more or less regular intervals, a column of heated water and steam. Old Faithful is one such geyser and is the most popular attraction of Yellowstone National Park, although it is not the largest or grandest geyser in the park. Old Faithful can vary in height from 100–180 feet with an average near 130–140 feet. Eruptions normally last between 1.5 to 5 minutes.

From August 1 to August 15, 1985, Old Faithful was observed and the waiting times between successive eruptions noted. There were 300 eruptions observed, so 299 waiting times were (in minutes) recorded and those shown in Table 7.1.

**Table 7.1**: `faithful` data (package *datasets*). Old Faithful geyser waiting times between two eruptions.

| waiting | waiting | waiting | waiting | waiting |
|---:|---:|---:|---:|---:|
| 79 | 83 | 75 | 76 | 50 |
| 54 | 71 | 59 | 63 | 82 |
| 74 | 64 | 89 | 88 | 54 |
| 62 | 77 | 79 | 52 | 75 |
| 85 | 81 | 59 | 93 | 78 |
| 55 | 59 | 81 | 49 | 79 |
| 88 | 84 | 50 | 57 | 78 |
| 85 | 48 | 85 | 77 | 78 |
| 51 | 82 | 59 | 68 | 70 |
| 85 | 60 | 87 | 81 | 79 |
| 54 | 92 | 53 | 81 | 70 |
| 84 | 78 | 69 | 73 | 54 |
| 78 | 78 | 77 | 50 | 86 |
| 47 | 65 | 56 | 85 | 50 |
| 83 | 73 | 88 | 74 | 90 |
| 52 | 82 | 81 | 55 | 54 |
| 62 | 56 | 45 | 77 | 54 |
| 84 | 79 | 82 | 83 | 77 |
| 52 | 71 | 55 | 83 | 79 |

**Table 7.1**: `faithful` data (continued).

| `waiting` | `waiting` | `waiting` | `waiting` | `waiting` |
|---:|---:|---:|---:|---:|
| 79 | 62 | 90 | 51 | 64 |
| 51 | 76 | 45 | 78 | 75 |
| 47 | 60 | 83 | 84 | 47 |
| 78 | 78 | 56 | 46 | 86 |
| 69 | 76 | 89 | 83 | 63 |
| 74 | 83 | 46 | 55 | 85 |
| 83 | 75 | 82 | 81 | 82 |
| 55 | 82 | 51 | 57 | 57 |
| 76 | 70 | 86 | 76 | 82 |
| 78 | 65 | 53 | 84 | 67 |
| 79 | 73 | 79 | 77 | 74 |
| 73 | 88 | 81 | 81 | 54 |
| 77 | 76 | 60 | 87 | 83 |
| 66 | 80 | 82 | 77 | 73 |
| 80 | 48 | 77 | 51 | 73 |
| 74 | 86 | 76 | 78 | 88 |
| 52 | 60 | 59 | 60 | 80 |
| 48 | 90 | 80 | 82 | 71 |
| 80 | 50 | 49 | 91 | 83 |
| 59 | 78 | 96 | 53 | 56 |
| 90 | 63 | 53 | 78 | 79 |
| 80 | 72 | 77 | 46 | 78 |
| 58 | 84 | 77 | 77 | 84 |
| 84 | 75 | 65 | 84 | 58 |
| 58 | 51 | 81 | 49 | 83 |
| 73 | 82 | 71 | 83 | 43 |
| 83 | 62 | 70 | 71 | 60 |
| 64 | 88 | 81 | 80 | 75 |
| 53 | 49 | 93 | 49 | 81 |
| 82 | 83 | 53 | 75 | 46 |
| 59 | 81 | 89 | 64 | 90 |
| 75 | 47 | 45 | 76 | 46 |
| 90 | 84 | 86 | 53 | 74 |
| 54 | 52 | 58 | 94 | |
| 80 | 86 | 78 | 55 | |
| 54 | 81 | 66 | 76 | |

The Hertzsprung-Russell (H-R) diagram forms the basis of the theory of stellar evolution. The diagram is essentially a plot of the energy output of stars plotted against their surface temperature. Data from the H-R diagram of Star Cluster CYG OB1, calibrated according to Vanisma and De Greve (1972) are shown in Table 7.2 (from Hand et al., 1994).

**Table 7.2**: `CYGOB1` data. Energy output and surface temperature of Star Cluster CYG OB1.

| `logst` | `logli` | `logst` | `logli` | `logst` | `logli` |
|---|---|---|---|---|---|
| 4.37 | 5.23 | 4.23 | 3.94 | 4.45 | 5.22 |
| 4.56 | 5.74 | 4.42 | 4.18 | 3.49 | 6.29 |
| 4.26 | 4.93 | 4.23 | 4.18 | 4.23 | 4.34 |
| 4.56 | 5.74 | 3.49 | 5.89 | 4.62 | 5.62 |
| 4.30 | 5.19 | 4.29 | 4.38 | 4.53 | 5.10 |
| 4.46 | 5.46 | 4.29 | 4.22 | 4.45 | 5.22 |
| 3.84 | 4.65 | 4.42 | 4.42 | 4.53 | 5.18 |
| 4.57 | 5.27 | 4.49 | 4.85 | 4.43 | 5.57 |
| 4.26 | 5.57 | 4.38 | 5.02 | 4.38 | 4.62 |
| 4.37 | 5.12 | 4.42 | 4.66 | 4.45 | 5.06 |
| 3.49 | 5.73 | 4.29 | 4.66 | 4.50 | 5.34 |
| 4.43 | 5.45 | 4.38 | 4.90 | 4.45 | 5.34 |
| 4.48 | 5.42 | 4.22 | 4.39 | 4.55 | 5.54 |
| 4.01 | 4.05 | 3.48 | 6.05 | 4.45 | 4.98 |
| 4.29 | 4.26 | 4.38 | 4.42 | 4.42 | 4.50 |
| 4.42 | 4.58 | 4.56 | 5.10 | | |

## 7.2 Density Estimation

The goal of density estimation is to approximate the probability density function of a random variable (univariate or multivariate) given a sample of observations of the variable. Univariate histograms are a simple example of a density estimate; they are often used for two purposes, counting and displaying the distribution of a variable, but according to Wilkinson (1992), they are effective for neither. For bivariate data, two-dimensional histograms can be constructed, but for small and moderate sized data sets that is not of any real use for estimating the bivariate density function, simply because most of the 'boxes' in the histogram will contain too few observations, or if the number of boxes is reduced the resulting histogram will be too coarse a representation of the density function.

The density estimates provided by one- and two-dimensional histograms can be improved on in a number of ways. If, of course, we are willing to assume a particular form for the variable's distribution, for example, Gaussian, density estimation would be reduced to estimating the parameters of the assumed distribution. More commonly, however, we wish to allow the data to speak for themselves and so one of a variety of non-parametric estimation procedures that are now available might be used. Density estimation is covered in detail in several books, including Silverman (1986), Scott (1992), Wand and Jones (1995) and Simonoff (1996). One of the most popular class of procedures is the kernel density estimators, which we now briefly describe for univariate and bivariate data.

### *7.2.1 Kernel Density Estimators*

From the definition of a probability density, if the random $X$ has a density $f$,

$$f(x) = \lim_{h \to 0} \frac{1}{2h} \mathsf{P}(x - h < X < x + h). \tag{7.1}$$

For any given $h$ a naïve estimator of $\mathsf{P}(x - h < X < x + h)$ is the proportion of the observations $x_1, x_2, \dots, x_n$ falling in the interval $(x - h, x + h)$, that is

$$\hat{f}(x) = \frac{1}{2hn} \sum_{i=1}^{n} I(x_i \in (x - h, x + h)), \tag{7.2}$$

i.e., the number of $x_1, \dots, x_n$ falling in the interval $(x - h, x + h)$ divided by $2hn$. If we introduce a weight function $W$ given by

$$W(x) = \begin{cases} \frac{1}{2} & |x| < 1 \\ 0 & \text{else} \end{cases}$$

then the naïve estimator can be rewritten as

$$\hat{f}(x) = \frac{1}{n} \sum_{i=1}^{n} \frac{1}{h} W\left(\frac{x - x_i}{h}\right). \tag{7.3}$$

Unfortunately this estimator is not a continuous function and is not particularly satisfactory for practical density estimation. It does however lead naturally to the kernel estimator defined by

$$\hat{f}(x) = \frac{1}{hn} \sum_{i=1}^{n} K\left(\frac{x - x_i}{h}\right) \tag{7.4}$$

where $K$ is known as the *kernel function* and $h$ as the *bandwidth* or *smoothing parameter*. The kernel function must satisfy the condition

$$\int_{-\infty}^{\infty} K(x)dx = 1.$$

Usually, but not always, the kernel function will be a symmetric density function for example, the normal. Three commonly used kernel functions are

**rectangular:**

$$K(x) = \begin{cases} \frac{1}{2} & |x| < 1 \\ 0 & \text{else} \end{cases}$$

**triangular:**

$$K(x) = \begin{cases} 1 - |x| & |x| < 1 \\ 0 & \text{else} \end{cases}$$

**Gaussian:**

$$K(x) = \frac{1}{\sqrt{2\pi}} e^{-\frac{1}{2}x^2}$$

The three kernel functions are implemented in R as shown in lines 1–3 of Figure 7.1. For some grid x, the kernel functions are plotted using the R statements in lines 5–11 (Figure 7.1).

The kernel estimator $\hat{f}$ is a sum of 'bumps' placed at the observations. The kernel function determines the shape of the bumps while the window width $h$ determines their width. Figure 7.2 (redrawn from a similar plot in Silverman, 1986) shows the individual bumps $n^{-1}h^{-1}K((x-x_i)/h)$, as well as the estimate $\hat{f}$ obtained by adding them up for an artificial set of data points

```
R> x <- c(0, 1, 1.1, 1.5, 1.9, 2.8, 2.9, 3.5)
R> n <- length(x)
```

For a grid

```
R> xgrid <- seq(from = min(x) - 1, to = max(x) + 1, by = 0.01)
```

on the real line, we can compute the contribution of each measurement in x, with $h = 0.4$, by the Gaussian kernel (defined in Figure 7.1, line 3) as follows;

```
R> h <- 0.4
R> bumps <- sapply(x, function(a) gauss((xgrid - a)/h)/(n *
+      h))
```

A plot of the individual bumps and their sum, the kernel density estimate $\hat{f}$, is shown in Figure 7.2.

The kernel density estimator considered as a sum of 'bumps' centred at the observations has a simple extension to two dimensions (and similarly for more than two dimensions). The bivariate estimator for data $(x_1, y_1)$, $(x_2, y_2)$, …, $(x_n, y_n)$ is defined as

$$\hat{f}(x, y) = \frac{1}{nh_x h_y} \sum_{i=1}^{n} K\left(\frac{x - x_i}{h_x}, \frac{y - y_i}{h_y}\right). \tag{7.5}$$

In this estimator each coordinate direction has its own smoothing parameter $h_x$ and $h_y$. An alternative is to scale the data equally for both dimensions and use a single smoothing parameter.

```
R> rec <- function(x) (abs(x) < 1) * 0.5
R> tri <- function(x) (abs(x) < 1) * (1 - abs(x))
R> gauss <- function(x) 1/sqrt(2 * pi) * exp(-(x^2)/2)
R> x <- seq(from = -3, to = 3, by = 0.001)
R> plot(x, rec(x), type = "l", ylim = c(0, 1), lty = 1,
+      ylab = expression(K(x)))
R> lines(x, tri(x), lty = 2)
R> lines(x, gauss(x), lty = 3)
R> legend(-3, 0.8, legend = c("Rectangular", "Triangular",
+      "Gaussian"), lty = 1:3, title = "kernel functions",
+      bty = "n")
```

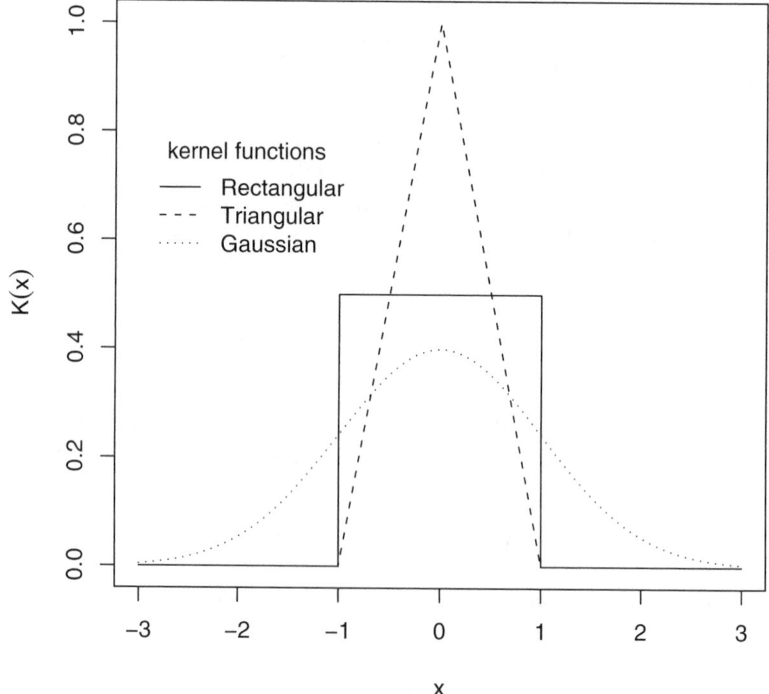


**Figure 7.1** Three commonly used kernel functions.

```
R> plot(xgrid, rowSums(bumps), ylab = expression(hat(f)(x)),
+      type = "l", xlab = "x", lwd = 2)
R> rug(x, lwd = 2)
R> out <- apply(bumps, 2, function(b) lines(xgrid,
+      b))
```

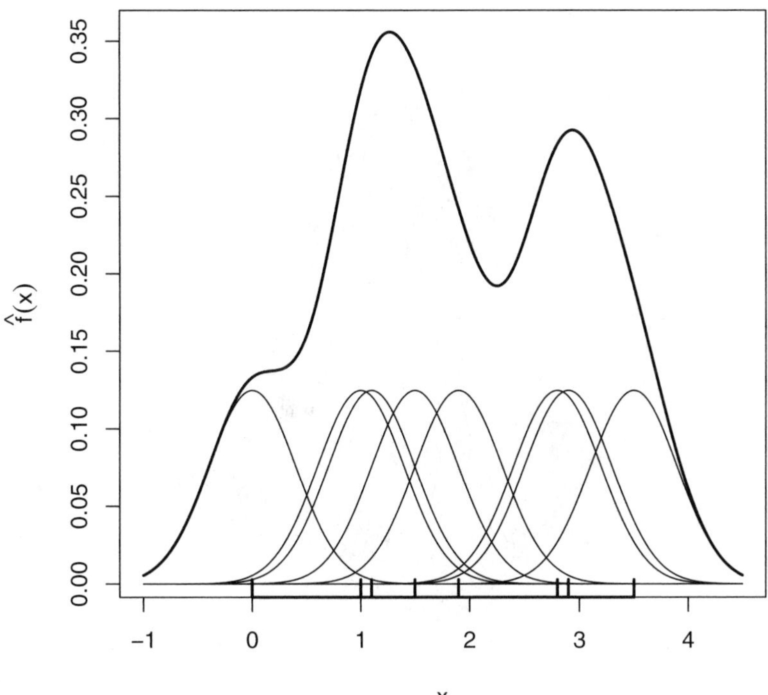


**Figure 7.2** Kernel estimate showing the contributions of Gaussian kernels evaluated for the individual observations with bandwidth $h = 0.4$.

For bivariate density estimation a commonly used kernel function is the standard bivariate normal density

$$K(x, y) = \frac{1}{2\pi} e^{-\frac{1}{2}(x^2+y^2)}.$$

Another possibility is the bivariate Epanechnikov kernel given by

$$K(x, y) = \begin{cases} \frac{2}{\pi}(1 - x^2 - y^2) & x^2 + y^2 < 1 \\ 0 & \text{else} \end{cases}$$

```
R> epa <- function(x, y) ((x^2 + y^2) < 1) * 2/pi *
+     (1 - x^2 - y^2)
R> x <- seq(from = -1.1, to = 1.1, by = 0.05)
R> epavals <- sapply(x, function(a) epa(a, x))
R> persp(x = x, y = x, z = epavals, xlab = "x", ylab = "y",
+     zlab = expression(K(x, y)), theta = -35, axes = TRUE,
+     box = TRUE)
```

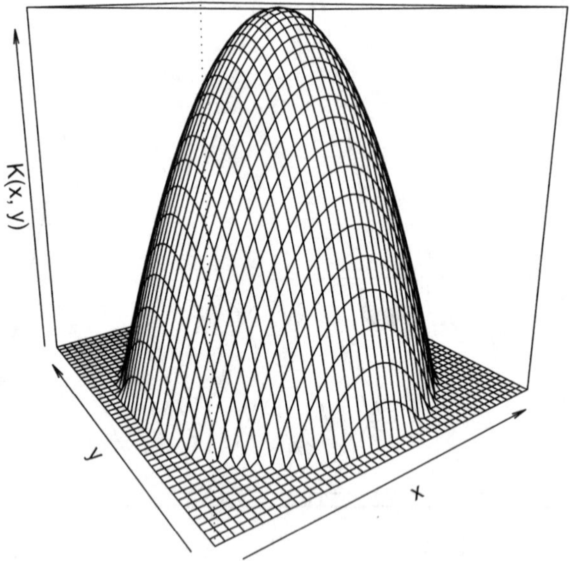


**Figure 7.3** Epanechnikov kernel for a grid between $(-1.1, -1.1)$ and $(1.1, 1.1)$.

which is implemented and depicted in Figure 7.3, here by using the `persp` function for plotting in three dimensions.

According to Venables and Ripley (2002) the bandwidth should be chosen to be proportional to $n^{-1/5}$; unfortunately the constant of proportionality depends on the unknown density. The tricky problem of bandwidth estimation is considered in detail in Silverman (1986).

## 7.3 Analysis Using R

The R function `density` can be used to calculate kernel density estimators with a variety of kernels (`window` argument). We can illustrate the function's use by applying it to the geyser data to calculate three density estimates of the data and plot each on a histogram of the data, using the code displayed with Figure 7.4. The `hist` function places an ordinary histogram of the geyser data in each of the three plotting regions (lines 4, 10, 17). Then, the `density` function with three different kernels (lines 8, 14, 21, with a Gaussian kernel being the default in line 8) is plotted in addition. The `rug` statement simply places the observations is vertical bars onto the x-axis. All three density estimates show that the waiting times between eruptions have a distinctly bimodal form, which we will investigate further in Subsection 7.3.1.

For the bivariate star data in Table 7.2 we can estimate the bivariate density using the `bkde2D` function from package *KernSmooth* (Wand and Ripley, 2005). The resulting estimate can then be displayed as a contour plot (using `contour`) or as a perspective plot (using `persp`). The resulting contour plot is shown in Figure 7.5, and the perspective plot in 7.6. Both clearly show the presence of two separated classes of stars.

### *7.3.1 A Parametric Density Estimate for the Old Faithful Data*

In the previous section we considered the non-parametric kernel density estimators for the Old Faithful data. The estimators showed the clear bimodality of the data and in this section this will be investigated further by fitting a parametric model based on a two-component normal mixture model. Such models are members of the class of finite mixture distributions described in great detail in McLachlan and Peel (2000). The two-component normal mixture distribution was first considered by Karl Pearson over 100 years ago (Pearson, 1894) and is given explicitly by

$$f(x) = p\phi(x, \mu_1, \sigma_1^2) + (1 - p)\phi(x, \mu_2, \sigma_2^2)$$

where $\phi(x, \mu, \sigma^2)$ denotes the normal density.

This distribution had five parameters to estimate, the mixing proportion, $p$, and the mean and variance of each component normal distribution. Pearson heroically attempted this by the method of moments, which required solving a polynomial equation of the $9^{\text{th}}$ degree. Nowadays the preferred estimation approach is maximum likelihood. The following R code contains a function to calculate the relevant log-likelihood and then uses the optimiser `optim` to find values of the five parameters that minimise the negative log-likelihood.

```
R> logL <- function(param, x) {
+     d1 <- dnorm(x, mean = param[2], sd = param[3])
+     d2 <- dnorm(x, mean = param[4], sd = param[5])
+     -sum(log(param[1] * d1 + (1 - param[1]) * d2))
+ }
```

```
R> data("faithful", package = "datasets")
R> x <- faithful$waiting
R> layout(matrix(1:3, ncol = 3))
R> hist(x, xlab = "Waiting times (in min.)",
+      ylab = "Frequency",
+      probability = TRUE, main = "Gaussian kernel",
+      border = "gray")
R> lines(density(x, width = 12), lwd = 2)
R> rug(x)
R> hist(x, xlab = "Waiting times (in min.)",
+      ylab = "Frequency",
+      probability = TRUE, main = "Rectangular kernel",
+      border = "gray")
R> lines(density(x, width = 12, window = "rectangular"),
+      lwd = 2)
R> rug(x)
R> hist(x, xlab = "Waiting times (in min.)",
+      ylab = "Frequency",
+      probability = TRUE, main = "Triangular kernel",
+      border = "gray")
R> lines(density(x, width = 12, window = "triangular"),
+      lwd = 2)
R> rug(x)
```

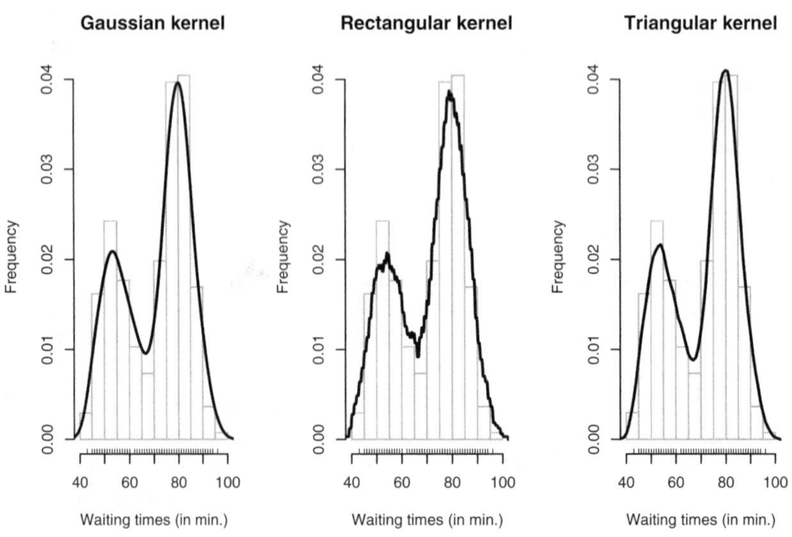


**Figure 7.4** Density estimates of the geyser eruption data imposed on a histogram of the data.

```
R> library("KernSmooth")
R> data("CYGOB1", package = "HSAUR")
R> CYGOB1d <- bkde2D(CYGOB1, bandwidth = sapply(CYGOB1,
+     dpik))
R> contour(x = CYGOB1d$x1, y = CYGOB1d$x2, z = CYGOB1d$fhat,
+     xlab = "log surface temperature",
+     ylab = "log light intensity")
```

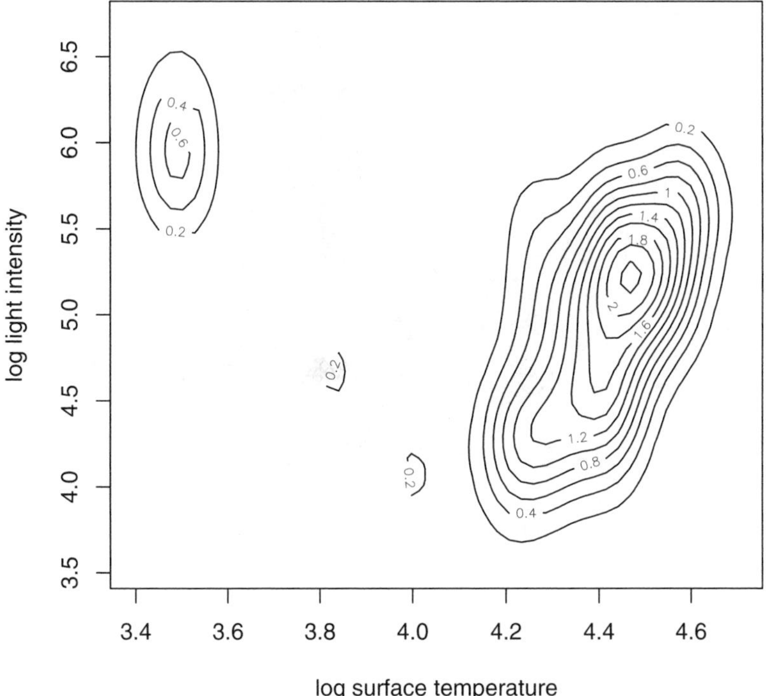


**Figure 7.5** A contour plot of the bivariate density estimate of the CYGOB1 data, i.e., a two-dimensional graphical display for a three-dimensional problem.

```
R> persp(x = CYGOB1d$x1, y = CYGOB1d$x2, z = CYGOB1d$fhat,
+      xlab = "log surface temperature",
+      ylab = "log light intensity",
+      zlab = "estimated density", theta = -35, axes = TRUE,
+      box = TRUE)
```

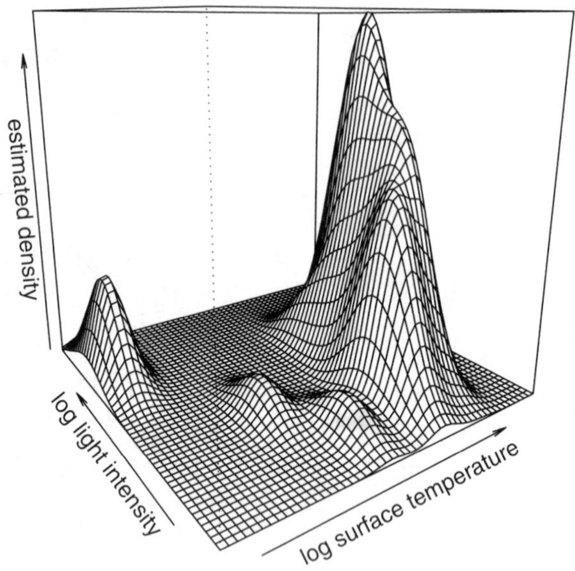


**Figure 7.6** The bivariate density estimate of the CYGOB1 data, here shown in a three-dimensional fashion using the persp function.

```
R> startparam <- c(p = 0.5, mu1 = 50, sd1 = 3, mu2 = 80,
+      sd2 = 3)
R> opp <- optim(startparam, logL, x = faithful$waiting,
+      method = "L-BFGS-B", lower = c(0.01, rep(1,
+          4)), upper = c(0.99, rep(200, 4)))
R> opp

$par
          p         mu1         sd1         mu2         sd2
```

```
 0.3608905 54.6120933  5.8723821 80.0934226  5.8672998

$value
[1] 1034.002

$counts
function gradient 
      55       55 

$convergence
[1] 0
```

Of course, optimising the appropriate likelihood 'by hand' is not very convenient. In fact, (at least) two packages offer high-level functionality for estimating mixture models. The first one is package *mclust* (Fraley et al., 2005) implementing the methodology described in Fraley and Raftery (2002). Here, a Bayesian information criterion (BIC) is applied to choose the form of the mixture model:

```
R> library("mclust")
R> mc <- Mclust(faithful$waiting)
R> mc
```

```
 best model: equal variance with 2 groups

 averge/median classification uncertainty: 0.015 / 0
```

and the estimated means are

```
R> mc$mu
```

```
       1        2 
80.09624 54.62190 
```

with estimated standard deviation (found to be equal within both groups)

```
R> sqrt(mc$sigmasq)
```

```
[1] 5.867345
```

The proportion is $\hat{p} = 0.64$. The second package is called *flexmix* whose functionality is described by Leisch (2004). A mixture of two normals can be fitted using

```
R> library("flexmix")
R> fl <- flexmix(waiting ~ 1, data = faithful, k = 2)
```

with $\hat{p} = 0.36$ and estimated parameters

```
R> parameters(fl, component = 1)
```

```
$coef
(Intercept) 
    54.6287 

$sigma
[1] 5.895234
```

```
R> parameters(fl, component = 2)

$coef
(Intercept)
   80.09858

$sigma
[1] 5.871749
```

The results are identical for all practical purposes and we can plot the fitted mixture and a single fitted normal into a histogram of the data using the R code which produces Figure 7.7. The `dnorm` function can be used to evaluate the normal density with given mean and standard deviation, here as estimated for the two-components of our mixture model, which are then collapsed into our density estimate `f`. Clearly the two-component mixture is a far better fit than a single normal distribution for these data.

We can get standard errors for the five parameter estimates by using a bootstrap approach (see Efron and Tibshirani, 1993). The original data are slightly perturbed by drawing $n$ out of $n$ observations *with replacement* and those artificial replications of the original data are called *bootstrap samples*. Now, we can fit the mixture for each bootstrap sample and assess the variability of the estimates, for example using confidence intervals. Some suitable R code based on the `Mclust` function follows. First, we define a function that, for a bootstrap sample `indx`, fits a two-component mixture model and returns $\hat{p}$ and the estimated means (note that we need to make sure that we always get an estimate of $p$, not $1 - p$):

```
R> library("boot")
R> fit <- function(x, indx) {
+      a <- Mclust(x[indx], minG = 2, maxG = 2)
+      if (a$pro[1] < 0.5)
+          return(c(p = a$pro[1], mu1 = a$mu[1], mu2 = a$mu[2]))
+      return(c(p = 1 - a$pro[1], mu1 = a$mu[2], mu2 = a$mu[1]))
+ }
```

The function `fit` can now be fed into the `boot` function (Canty and Ripley, 2005) for bootstrapping (here 1000 bootstrap samples are drawn)

```
R> bootpara <- boot(faithful$waiting, fit, R = 1000)
```

We assess the variability of our estimates $\hat{p}$ by means of adjusted bootstrap percentile (BCa) confidence intervals, which for $\hat{p}$ can be obtained from

```
R> boot.ci(bootpara, type = "bca", index = 1)

BOOTSTRAP CONFIDENCE INTERVAL CALCULATIONS
Based on 1000 bootstrap replicates

CALL :
boot.ci(boot.out = bootpara, type = "bca", index = 1)

Intervals :
```

```
R> opar <- as.list(opp$par)
R> rx <- seq(from = 40, to = 110, by = 0.1)
R> d1 <- dnorm(rx, mean = opar$mu1, sd = opar$sd1)
R> d2 <- dnorm(rx, mean = opar$mu2, sd = opar$sd2)
R> f <- opar$p * d1 + (1 - opar$p) * d2
R> hist(x, probability = TRUE,
+     xlab = "Waiting times (in min.)",
+     border = "gray", xlim = range(rx), ylim = c(0,
+         0.06), main = "")
R> lines(rx, f, lwd = 2)
R> lines(rx, dnorm(rx, mean = mean(x), sd = sd(x)),
+     lty = 2, lwd = 2)
R> legend(50, 0.06,
+     legend = c("Fitted two-component mixture density",
+     "Fitted single normal density"), lty = 1:2,
+     bty = "n")
```

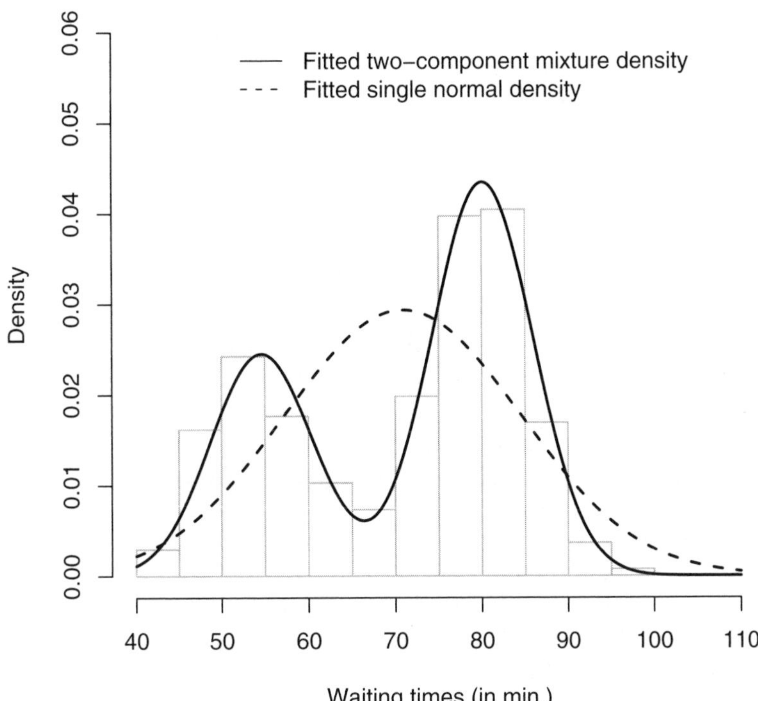


**Figure 7.7** Fitted normal density and two-component normal mixture for geyser eruption data.

```
Level        BCa
95%    ( 0.3041,  0.4233 )
Calculations and Intervals on Original Scale
```

We see that there is a reasonable variability in the mixture model, however, the means in the two components are rather stable, as can be seen from

```
R> boot.ci(bootpara, type = "bca", index = 2)
```

```
BOOTSTRAP CONFIDENCE INTERVAL CALCULATIONS
Based on 1000 bootstrap replicates

CALL :
boot.ci(boot.out = bootpara, type = "bca", index = 2)

Intervals :
Level        BCa
95%    (53.42, 56.07 )
Calculations and Intervals on Original Scale
```

for $\hat{\mu}_1$ and for $\hat{\mu}_2$ from

```
R> boot.ci(bootpara, type = "bca", index = 3)
```

```
BOOTSTRAP CONFIDENCE INTERVAL CALCULATIONS
Based on 1000 bootstrap replicates

CALL :
boot.ci(boot.out = bootpara, type = "bca", index = 3)

Intervals :
Level        BCa
95%    (79.05, 81.01 )
Calculations and Intervals on Original Scale
```

Finally, we show a graphical representation of both the bootstrap distribution of the mean estimates *and* the corresponding confidence intervals. For convenience, we define a function for plotting, namely

```
R> bootplot <- function(b, index, main = "") {
+     dens <- density(b$t[, index])
+     ci <- boot.ci(b, type = "bca", index = index)$bca[4:5]
+     est <- b$t0[index]
+     plot(dens, main = main)
+     y <- max(dens$y)/10
+     segments(ci[1], y, ci[2], y, lty = 2)
+     points(ci[1], y, pch = "(")
+     points(ci[2], y, pch = ")")
+     points(est, y, pch = 19)
+ }
```

The element `t` of an object created by `boot` contains the bootstrap replications of our estimates, i.e., the values computed by `fit` for each of the 1000

```
R> layout(matrix(1:2, ncol = 2))
R> bootplot(bootpara, 2, main = expression(mu[1]))
R> bootplot(bootpara, 3, main = expression(mu[2]))
```

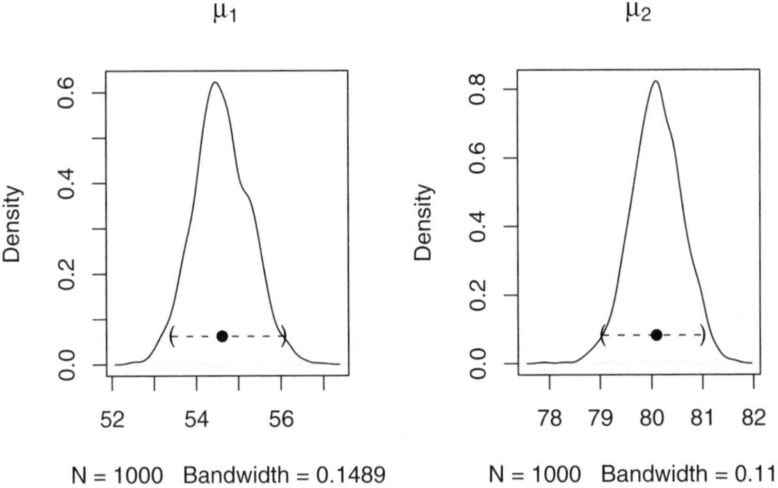


**Figure 7.8** Bootstrap distribution and confidence intervals for the mean estimates of a two-component mixture for the geyser data.

bootstrap samples of the geyser data. First, we plot a simple density estimate and then construct a line representing the confidence interval. We apply this function to the bootstrap distributions of our estimates $\hat{\mu}_1$ and $\hat{\mu}_2$ in Figure 7.8.

## 7.4 Summary

Histograms and scatterplots are frequently used to give graphical representations of univariate and bivariate data. But both can often be improved and made more helpful by adding some form of density estimate. For scatterplots in particular adding a contour plot of the estimated bivariate density can be particularly useful in aiding in the identification of clusters, gaps and outliers.

### Exercises

Ex. 7.1 The data shown in Table 7.3 are the velocities of 82 galaxies from six well-separated conic sections of space (Postman et al., 1986, Roeder, 1990). The data are intended to shed light on whether or not the observable universe contains superclusters of galaxies surrounded by large voids. The evidence for the existence of superclusters would be the multimodality of the distribution of velocities. Construct a histogram of the data and add a variety of kernel estimates of the density function. What do you conclude about the possible existence of superclusters of galaxies?

**Table 7.3**: `galaxies` data (package *MASS*). Velocities of 82 galaxies.

| galaxies | galaxies | galaxies | galaxies | galaxies |
|---|---|---|---|---|
| 9172 | 19349 | 20196 | 22209 | 23706 |
| 9350 | 19440 | 20215 | 22242 | 23711 |
| 9483 | 19473 | 20221 | 22249 | 24129 |
| 9558 | 19529 | 20415 | 22314 | 24285 |
| 9775 | 19541 | 20629 | 22374 | 24289 |
| 10227 | 19547 | 20795 | 22495 | 24366 |
| 10406 | 19663 | 20821 | 22746 | 24717 |
| 16084 | 19846 | 20846 | 22747 | 24990 |
| 16170 | 19856 | 20875 | 22888 | 25633 |
| 18419 | 19863 | 20986 | 22914 | 26690 |
| 18552 | 19914 | 21137 | 23206 | 26995 |
| 18600 | 19918 | 21492 | 23241 | 32065 |
| 18927 | 19973 | 21701 | 23263 | 32789 |
| 19052 | 19989 | 21814 | 23484 | 34279 |
| 19070 | 20166 | 21921 | 23538 | |
| 19330 | 20175 | 21960 | 23542 | |
| 19343 | 20179 | 22185 | 23666 | |

*Source*: From Roeder, K., *J. Am. Stat. Assoc.*, 85, 617–624, 1990. Reprinted with permission from *The Journal of the American Statistical Association*. 

Ex. 7.2 The data in Table 7.4 give the birth and death rates for 69 countries (from Hartigan, 1975). Produce a scatterplot of the data that shows a contour plot of the estimated bivariate density. Does the plot give you any interesting insights into the possible structure of the data?

**Table 7.4**: `birthdeathrates` data. Birth and death rates for 69 countries.

| birth | death | birth | death | birth | death |
|---|---|---|---|---|---|
| 36.4 | 14.6 | 26.2 | 4.3 | 18.2 | 12.2 |
| 37.3 | 8.0 | 34.8 | 7.9 | 16.4 | 8.2 |
| 42.1 | 15.3 | 23.4 | 5.1 | 16.9 | 9.5 |
| 55.8 | 25.6 | 24.8 | 7.8 | 17.6 | 19.8 |
| 56.1 | 33.1 | 49.9 | 8.5 | 18.1 | 9.2 |
| 41.8 | 15.8 | 33.0 | 8.4 | 18.2 | 11.7 |
| 46.1 | 18.7 | 47.7 | 17.3 | 18.0 | 12.5 |
| 41.7 | 10.1 | 46.6 | 9.7 | 17.4 | 7.8 |
| 41.4 | 19.7 | 45.1 | 10.5 | 13.1 | 9.9 |
| 35.8 | 8.5 | 42.9 | 7.1 | 22.3 | 11.9 |
| 34.0 | 11.0 | 40.1 | 8.0 | 19.0 | 10.2 |
| 36.3 | 6.1 | 21.7 | 9.6 | 20.9 | 8.0 |
| 32.1 | 5.5 | 21.8 | 8.1 | 17.5 | 10.0 |
| 20.9 | 8.8 | 17.4 | 5.8 | 19.0 | 7.5 |
| 27.7 | 10.2 | 45.0 | 13.5 | 23.5 | 10.8 |
| 20.5 | 3.9 | 33.6 | 11.8 | 15.7 | 8.3 |
| 25.0 | 6.2 | 44.0 | 11.7 | 21.5 | 9.1 |
| 17.3 | 7.0 | 44.2 | 13.5 | 14.8 | 10.1 |
| 46.3 | 6.4 | 27.7 | 8.2 | 18.9 | 9.6 |
| 14.8 | 5.7 | 22.5 | 7.8 | 21.2 | 7.2 |
| 33.5 | 6.4 | 42.8 | 6.7 | 21.4 | 8.9 |
| 39.2 | 11.2 | 18.8 | 12.8 | 21.6 | 8.7 |
| 28.4 | 7.1 | 17.1 | 12.7 | 25.5 | 8.8 |

*Source*: From Hartigan, J. A., *Clustering Algorithms*, Wiley, New York, 1975. With permission.

Ex. 7.3 A sex difference in the age of onset of schizophrenia was noted by Kraepelin (1919). Subsequent epidemiological studies of the disorder have consistently shown an earlier onset in men than in women. One model that has been suggested to explain this observed difference is known as the *subtype model* which postulates two types of schizophrenia, one characterised by early onset, typical symptoms and poor premorbid competence, and the other by late onset, atypical symptoms and good premorbid competence. The early onset type is assumed to be largely a disorder of men and the late onset largely a disorder of women. By fitting finite mixtures of normal densities separately to the onset data for men and women given in Table 7.5 see if you can produce some evidence for or against the subtype model.

**Table 7.5:** `schizophrenia` data. Age on onset of schizophrenia for both sexes.

| age | gender | age | gender | age | gender | age | gender |
|---|---|---|---|---|---|---|---|
| 20 | female | 20 | female | 22 | male | 27 | male |
| 30 | female | 43 | female | 19 | male | 18 | male |
| 21 | female | 39 | female | 16 | male | 43 | male |
| 23 | female | 40 | female | 16 | male | 20 | male |
| 30 | female | 26 | female | 18 | male | 17 | male |
| 25 | female | 50 | female | 16 | male | 21 | male |
| 13 | female | 17 | female | 33 | male | 5 | male |
| 19 | female | 17 | female | 22 | male | 27 | male |
| 16 | female | 23 | female | 23 | male | 25 | male |
| 25 | female | 44 | female | 10 | male | 18 | male |
| 20 | female | 30 | female | 14 | male | 24 | male |
| 25 | female | 35 | female | 15 | male | 33 | male |
| 27 | female | 20 | female | 20 | male | 32 | male |
| 43 | female | 41 | female | 11 | male | 29 | male |
| 6 | female | 18 | female | 25 | male | 34 | male |
| 21 | female | 39 | female | 9 | male | 20 | male |
| 15 | female | 27 | female | 22 | male | 21 | male |
| 26 | female | 28 | female | 25 | male | 31 | male |
| 23 | female | 30 | female | 20 | male | 22 | male |
| 21 | female | 34 | female | 19 | male | 15 | male |
| 23 | female | 33 | female | 22 | male | 27 | male |
| 23 | female | 30 | female | 23 | male | 26 | male |
| 34 | female | 29 | female | 24 | male | 23 | male |
| 14 | female | 46 | female | 29 | male | 47 | male |
| 17 | female | 36 | female | 24 | male | 17 | male |
| 18 | female | 58 | female | 22 | male | 21 | male |
| 21 | female | 28 | female | 26 | male | 16 | male |
| 16 | female | 30 | female | 20 | male | 21 | male |
| 35 | female | 28 | female | 25 | male | 19 | male |
| 32 | female | 37 | female | 17 | male | 31 | male |
| 48 | female | 31 | female | 25 | male | 34 | male |
| 53 | female | 29 | female | 28 | male | 23 | male |
| 51 | female | 32 | female | 22 | male | 23 | male |
| 48 | female | 48 | female | 22 | male | 20 | male |
| 29 | female | 49 | female | 23 | male | 21 | male |
| 25 | female | 30 | female | 35 | male | 18 | male |
| 44 | female | 21 | male | 16 | male | 26 | male |
| 23 | female | 18 | male | 29 | male | 30 | male |
| 36 | female | 23 | male | 33 | male | 17 | male |
| 58 | female | 21 | male | 15 | male | 21 | male |
| 28 | female | 27 | male | 29 | male | 19 | male |

**Table 7.5**: `schizophrenia` data (continued).

| age | gender | age | gender | age | gender | age | gender |
|---|---|---|---|---|---|---|---|
| 51 | female | 24 | male | 20 | male | 22 | male |
| 40 | female | 20 | male | 29 | male | 52 | male |
| 43 | female | 12 | male | 24 | male | 19 | male |
| 21 | female | 15 | male | 39 | male | 24 | male |
| 48 | female | 19 | male | 10 | male | 19 | male |
| 17 | female | 21 | male | 20 | male | 19 | male |
| 23 | female | 22 | male | 23 | male | 33 | male |
| 28 | female | 19 | male | 15 | male | 32 | male |
| 44 | female | 24 | male | 18 | male | 29 | male |
| 28 | female | 9 | male | 20 | male | 58 | male |
| 21 | female | 19 | male | 21 | male | 39 | male |
| 31 | female | 18 | male | 30 | male | 42 | male |
| 22 | female | 17 | male | 21 | male | 32 | male |
| 56 | female | 23 | male | 18 | male | 32 | male |
| 60 | female | 17 | male | 19 | male | 46 | male |
| 15 | female | 23 | male | 15 | male | 38 | male |
| 21 | female | 19 | male | 19 | male | 44 | male |
| 30 | female | 37 | male | 18 | male | 35 | male |
| 26 | female | 26 | male | 25 | male | 45 | male |
| 28 | female | 22 | male | 17 | male | 41 | male |
| 23 | female | 24 | male | 15 | male | 31 | male |
| 21 | female | 19 | male | 42 | male | | |

CHAPTER 8

# Recursive Partitioning: Large Companies and Glaucoma Diagnosis

## 8.1 Introduction

The Forbes 2000 list of the world's biggest industrial companies was introduced in detail in Chapter 1. Here, our interest is to construct a model explaining the profit of a company based on assets, sales and the market value.

A second set of data that will also be used in this chapter involves the investigation reported in Mardin et al. (2003) of whether laser scanner images of the eye background can be used to classify a patient's eye as suffering from glaucoma or not. Glaucoma is a neuro-degenerative disease of the optic nerve and is one of the major reasons for blindness in elderly people. For 196 people, 98 patients suffering glaucoma and 98 controls which have been matched by age and sex, 62 numeric variables derived from the laser scanning images are available. The data are available as `GlaucomaM` from package *ipred* (Peters et al., 2002). The variables describe the morphology of the optic nerve head, i.e., measures of volumes and areas in certain regions of the eye background. Those regions have been manually outlined by a physician. Our aim is to construct a prediction model which is able to decide whether an eye is affected by glaucomateous changes based on the laser image data.

Both sets of data described above could be analysed using the regression models described in Chapters 5 and 6, i.e., regression models for numeric and binary response variables based on a linear combination of the covariates. But here we shall employ an alternative approach known as *recursive partitioning*, where the resulting models are usually called *regression or classification trees*. This method was originally invented to deal with possible non-linear relationships between covariates and response. The basic idea is to partition the covariate space and to compute simple statistics of the dependent variable, like the mean or median, inside each cell.

## 8.2 Recursive Partitioning

There exist many algorithms for the construction of classification or regression trees but the majority of algorithms follow a simple general rule: First partition the observations by univariate splits in a recursive way and second fit a constant model in each cell of the resulting partition. An overview of this field of regression models is given by Murthy (1998)

In more details, for the first step, one selects a covariate $x_j$ from the $q$

available covariates $x_1, \dots, x_q$ and estimates a split point which separates the response values $y_i$ into two groups. For an ordered covariate $x_j$ a split point is a number $\xi$ dividing the observations into two groups. The first group consists of all observations with $x_j \le \xi$ and the second group contains the observations satisfying $x_j > \xi$. For a nominal covariate $x_j$, the two groups are defined by a set of levels $A$ where either $x_j \in A$ or $x_j \notin A$.

Once that the splits $\xi$ or $A$ for some selected covariate $x_j$ have been estimated, one applies the procedure sketched above for all observations in the first group and, recursively, splits this set of observations further. The same happens for all observations in the second group. The recursion is stopped when some stopping criterion is fulfilled.

The available algorithms mostly differ with respect to three points: how the covariate is selected in each step, how the split point is estimated and which stopping criterion is applied. One of the most popular algorithms is described in the 'Classification and Regression Trees' book by Breiman et al. (1984) and is available in R by the functions in package *rpart* (Therneau and Atkinson, 1997, Therneau et al., 2005). This algorithm first examines all possible splits for all covariates and chooses the split which leads to two groups that are 'purer' than the current group with respect to the values of the response variable $y$. There are many possible measures of impurity available, for regression problems with nominal response the *Gini* criterion is the default in *rpart*, alternatives and a more detailed description of tree based methods can be found in Ripley (1996).

The question when the recursion needs to stop is all but trivial. In fact, trees with too many leaves will suffer from overfitting and small trees will miss important aspects of the problem. Commonly, this problem is addressed by so-called *pruning* methods. As the name suggests, one first grows a very large tree using a trivial stopping criterion as the number of observations in a leaf, say, and then prunes branches that are not necessary.

Once that a tree has been grown, a simple summary statistic is computed for each leaf. The mean or median can be used for continuous responses and for nominal responses the proportions of the classes is commonly used. The prediction of a new observation is simply the corresponding summary statistic of the leaf to which this observation belongs.

However, even the right-sized tree consists of binary splits which are, of course, hard decisions. When the underlying relationship between covariate and response is smooth, such a split point estimate will be affected by high variability. This problem is addressed by so called *ensemble methods*. Here, multiple trees are grown on perturbed instances of the data set and their predictions are averaged. The simplest representative of such a procedure is called *bagging* (Breiman, 1996) and works as follows. We draw $B$ bootstrap samples from the original data set, i.e., we draw $n$ out of $n$ observations with replacement from our $n$ original observations. For each of those bootstrap samples we grow a very large tree. When we are interested in the prediction for a new observation, we pass this observation through all $B$ trees and average their predictions. It has been shown that the goodness of the predictions for future cases can be improved dramatically by this or similar simple procedures. More details can be found in Bühlmann (2004).

## 8.3 Analysis Using R

### 8.3.1 Forbes 2000 Data

For all companies from the Forbes 2000 list the assets, the market value and the sales are available. We expect those variables to have some impact on the profit and our aim is to explore this relationship for this data set of 2000 companies. Of course, the Forbes 2000 list is not a random sample of independent observations and thus a rather informal method such as a regression tree is appropriate to extract informations about simple relationships.

For some observations the profit is missing and we first remove those companies from the list

```
R> data("Forbes2000", package = "HSAUR")
R> Forbes2000 <- subset(Forbes2000, !is.na(profits))
```

The `rpart` function from *rpart* can be used to grow a regression tree. The response variable and the covariates are defined by a model formula in the same way as for `lm`, say. By default, a large initial tree is grown.

```
R> library("rpart")
R> forbes_rpart <- rpart(profits ~ assets + marketvalue +
+     sales, data = Forbes2000)
```

A `print` method for *rpart* objects is available, however, a graphical representation shown in Figure 8.1 is more convenient. Observations which satisfy the condition shown for each node go to the left and observations which don't are element of the right branch in each node. The numbers plotted in the leaves are the mean profit for those observations satisfying the conditions stated above. For example, the highest profit is observed for companies with a market value greater than 89.33 billion US dollars and with more than 91.92 US dollars sales.

To determine if the tree is appropriate or if some of the branches need to be subjected to pruning we can use the `cptable` element of the *rpart* object:

```
R> print(forbes_rpart$cptable)
```

```
          CP nsplit rel error    xerror      xstd
1 0.23748446      0 1.0000000 1.0010339 0.1946331
2 0.04600397      1 0.7625155 0.8397144 0.2174245
3 0.04258786      2 0.7165116 0.8066685 0.2166339
4 0.02030891      3 0.6739237 0.7625940 0.2089684
5 0.01854336      4 0.6536148 0.7842574 0.2093683
6 0.01102304      5 0.6350714 0.7925891 0.2106088
7 0.01076006      6 0.6240484 0.7931405 0.2128048
8 0.01000000      7 0.6132883 0.7902771 0.2128037
```

```
R> opt <- which.min(forbes_rpart$cptable[, "xerror"])
```

```
R> plot(forbes_rpart, uniform = TRUE, margin = 0.1,
+      branch = 0.5, compress = TRUE)
R> text(forbes_rpart)
```

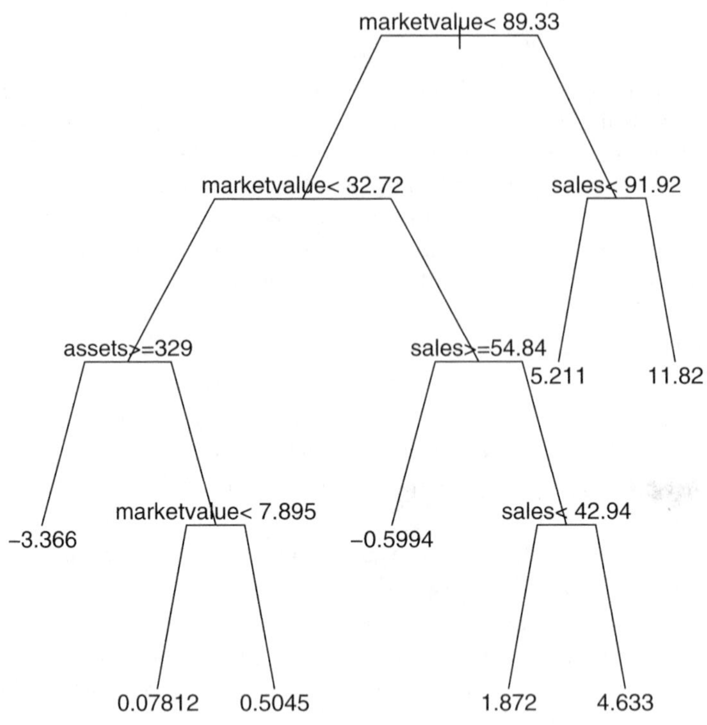


**Figure 8.1** Large initial tree for Forbes 2000 data.

The `xerror` column contains of estimates of cross-validated prediction error for different numbers of splits (`nsplit`). The best tree has three splits. Now we can prune back the large initial tree using

```
R> cp <- forbes_rpart$cptable[opt, "CP"]
R> forbes_prune <- prune(forbes_rpart, cp = cp)
```

The result is shown in Figure 8.2. This tree is much smaller. From the sample sizes and boxplots shown for each leaf we see that the majority of companies is grouped together. However, a large market value, more that 32.72 billion US dollars, seems to be a good indicator of large profits.

```
R> layout(matrix(1:2, nc = 1))
R> plot(forbes_prune, uniform = TRUE, margin = 0.1,
+     branch = 0.5, compress = TRUE)
R> text(forbes_prune)
R> rn <- rownames(forbes_prune$frame)
R> lev <- rn[sort(unique(forbes_prune$where))]
R> where <- factor(rn[forbes_prune$where], levels = lev)
R> n <- tapply(Forbes2000$profits, where, length)
R> boxplot(Forbes2000$profits ~ where, varwidth = TRUE,
+     ylim = range(Forbes2000$profit) * 1.3,
+     pars = list(axes = FALSE),
+     ylab = "Profits in US dollars")
R> abline(h = 0, lty = 3)
R> axis(2)
R> text(1:length(n), max(Forbes2000$profit) * 1.2,
+     paste("n = ", n))
```

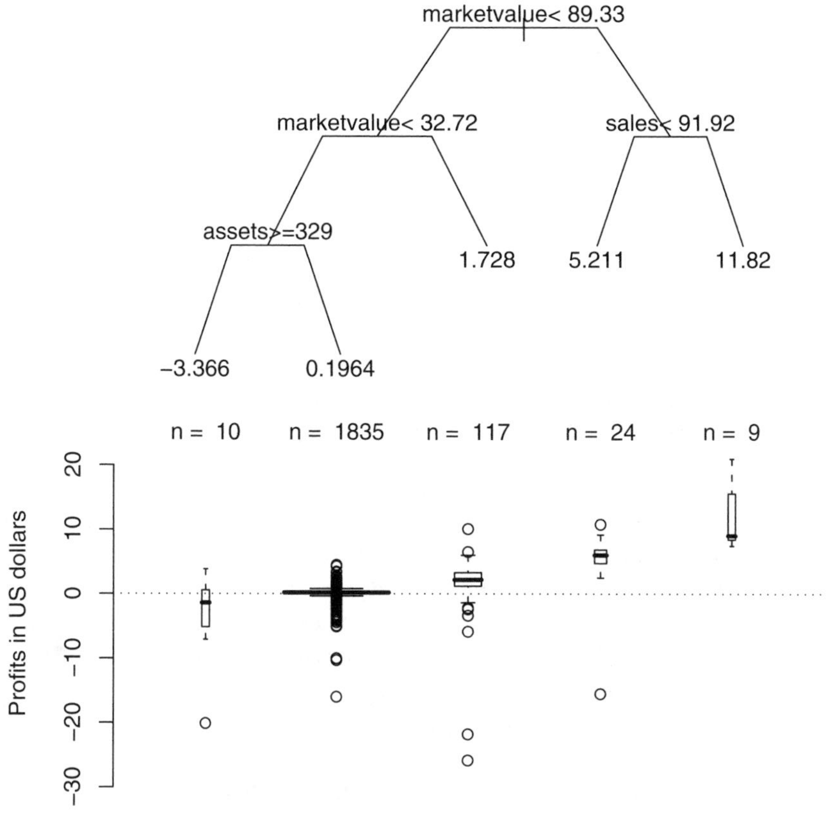


**Figure 8.2** Pruned regression tree for Forbes 2000 data with the distribution of the profit in each leaf depicted by a boxplot.

### 8.3.2 Glaucoma Diagnosis

The motivation for the analysis of the Forbes 2000 data by means of a regression tree was our aim to explore the relationship between the covariates and the response using a simple model and thus the graphical representation in Figure 8.2 is a good summary of the properties of the 2000 companies. For the glaucoma data we are primarily interested in the construction of a predictor. The relationship between the 62 covariates and the glaucoma status itself is not very interesting.

We start with a large initial tree and prune back branches according to the cross-validation criterion. The default is to use 10 runs of 10-fold cross-validation and we choose 100 runs of 10-fold cross-validation for reasons to be explained later.

```
R> data("GlaucomaM", package = "ipred")
R> glaucoma_rpart <- rpart(Class ~ ., data = GlaucomaM,
+     control = rpart.control(xval = 100))
R> glaucoma_rpart$cptable
```

```
          CP nsplit rel error    xerror       xstd
1 0.65306122      0 1.0000000 1.5306122 0.06054391
2 0.07142857      1 0.3469388 0.3877551 0.05647630
3 0.01360544      2 0.2755102 0.3775510 0.05590431
4 0.01000000      5 0.2346939 0.4489796 0.05960655
```

```
R> opt <- which.min(glaucoma_rpart$cptable[, "xerror"])
R> cp <- glaucoma_rpart$cptable[opt, "CP"]
R> glaucoma_prune <- prune(glaucoma_rpart, cp = cp)
```

The pruned tree consists of three leaves only (Figure 8.3), the class distribution in each leaf is depicted using a mosaicplot as produced by `mosaicplot`. For most eyes, the decision about the disease is based on the variable `varg`, a measurement of the volume of the optic nerve above some reference plane. A volume larger than 0.209 $mm^3$ indicates that the eye is healthy and a damage of the optic nerve head associated with loss of optic nerves (`varg` smaller than 0.209 $mm^3$) indicates a glaucomateous change.

As we discussed earlier, the choice of the appropriate sized tree is not a trivial problem. For the glaucoma data, the above choice of three leaves is very unstable across multiple runs of cross-validation. As an illustration of this problem we repeat the very same analysis as shown above and record the optimal number of splits as suggested by the cross-validation runs.

```
R> nsplitopt <- vector(mode = "integer", length = 25)
R> for (i in 1:length(nsplitopt)) {
+     cp <- rpart(Class ~ ., data = GlaucomaM)$cptable
+     nsplitopt[i] <- cp[which.min(cp[, "xerror"]),
+         "nsplit"]
+ }
R> table(nsplitopt)
```

```
R> layout(matrix(1:2, nc = 1))
R> plot(glaucoma_prune, uniform = TRUE, margin = 0.1,
+      branch = 0.5, compress = TRUE)
R> text(glaucoma_prune, use.n = TRUE)
R> rn <- rownames(glaucoma_prune$frame)
R> lev <- rn[sort(unique(glaucoma_prune$where))]
R> where <- factor(rn[glaucoma_prune$where], levels = lev)
R> mosaicplot(table(where, GlaucomaM$Class), main = "",
+      xlab = "", las = 1)
```

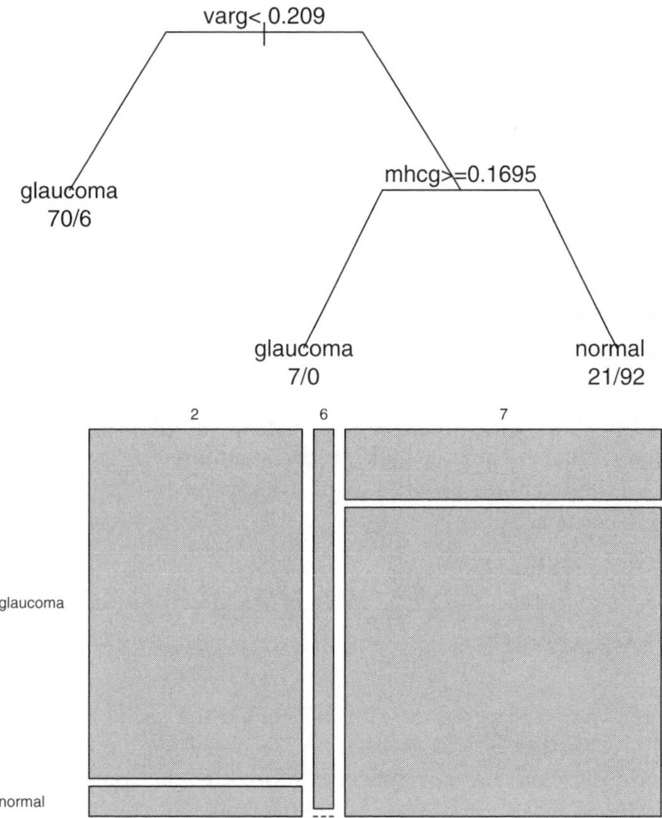


**Figure 8.3** Pruned classification tree of the glaucoma data with class distribution in the leaves depicted by a mosaicplot.

```
nsplitopt
 1  2  5
14  7  4
```

Although for 14 runs of cross-validation a simple tree with one split only is suggested, larger trees would have been favored in 11 of the cases. This short analysis shows that we should not trust the tree in Figure 8.3 too much.

One way out of this dilemma is the aggregation of multiple trees via *bagging*. In R, the bagging idea can be implemented by three or four lines of code. Case count or weight vectors representing the bootstrap samples can be drawn from the multinominal distribution with parameters $n$ and $p_1 = 1/n, \dots, p_n = 1/n$ via the `rmultinom` function. For each weight vector, one large tree is constructed without pruning and the *rpart* objects are stored in a list, here called `trees`:

```
R> trees <- vector(mode = "list", length = 25)
R> n <- nrow(GlaucomaM)
R> bootsamples <- rmultinom(length(trees), n, rep(1,
+     n)/n)
R> mod <- rpart(Class ~ ., data = GlaucomaM,
+     control = rpart.control(xval = 0))
R> for (i in 1:length(trees)) trees[[i]] <- update(mod,
+     weights = bootsamples[, i])
```

The `update` function re-evaluates the call of `mod`, however, with the weights being altered, i.e., fits a tree to a bootstrap sample specified by the weights. It is interesting to have a look at the structures of the multiple trees. For example, the variable selected for splitting in the root of the tree is not unique as can be seen by

```
R> table(sapply(trees,
+     function(x) as.character(x$frame$var[1])))

phcg varg vari vars
   1   14    9    1
```

Although `varg` is selected most of the time, other variables such as `vari` occur as well – a further indication that the tree in Figure 8.3 is questionable and that hard decisions are not appropriate for the glaucoma data.

In order to make use of the ensemble of trees in the list `trees` we estimate the conditional probability of suffering from glaucoma given the covariates for each observation in the original data set by

```
R> classprob <- matrix(0, nrow = n, ncol = length(trees))
R> for (i in 1:length(trees)) {
+     classprob[, i] <- predict(trees[[i]],
+     newdata = GlaucomaM)[,
+         2]
+     classprob[bootsamples[, i] > 0, i] <- NA
+ }
```

Thus, for each observation we get 25 estimates. However, each observation has been used for growing one of the trees with probability 0.632 and thus was not used with probability 0.368. Consequently, the estimate from a tree where an observation was not used for growing is better for judging the quality of the predictions and we label the other estimates with NA.

Now, we can average the estimates and we vote for glaucoma when the average of the estimates of the conditional glaucoma probability exceeds 0.5. The comparison between the observed and the predicted classes does not suffer from overfitting since the predictions are computed from those trees for which each single observation was *not* used for growing.

```
R> avg <- rowMeans(classprob, na.rm = TRUE)
R> predictions <- factor(avg > 0.5,
+     labels = levels(GlaucomaM$Class))
R> predtab <- table(predictions, GlaucomaM$Class)
R> predtab
```

```
predictions glaucoma normal
   glaucoma       76     14
   normal         22     84
```

Thus, an honest estimate of the probability of a glaucoma prediction when the patient is actually suffering from glaucoma is

```
R> round(predtab[1, 1]/colSums(predtab)[1] * 100)
```

```
glaucoma
      78
```

per cent. For

```
R> round(predtab[2, 2]/colSums(predtab)[2] * 100)
```

```
normal
    86
```

per cent of normal eyes, the ensemble does not predict a glaucomateous damage.

Although we are mainly interested in a predictor, i.e., a *black box* machine for predicting glaucoma is our main focus, the nature of the black box might be interesting as well. From the classification tree analysis shown above we expect to see a relationship between the volume above the reference plane (varg) and the estimated conditional probability of suffering from glaucoma. A graphical approach is sufficient here and we simply plot the observed values of varg against the averages of the estimated glaucoma probability (such plots have been used by Breiman, 2001b, Garczarek and Weihs, 2003, for example). In addition, we construct such a plot for another covariate as well, namely vari, the volume above the reference plane measured in the inferior part of the optic nerve head only. Figure 8.4 shows that the initial split of $0.209mm^3$ for varg (see Figure 8.3) corresponds to the ensemble predictions rather well.

The *bagging* procedure is a special case of a more general approach called *random forest* (Breiman, 2001a). The package *randomForest* (Breiman et al., 2005) can be used to compute such ensembles via

```
R> library("lattice")
R> gdata <- data.frame(avg = rep(avg, 2),
+     class = rep(as.numeric(GlaucomaM$Class),
+     2), obs = c(GlaucomaM[["varg"]], GlaucomaM[["vari"]]),
+     var = factor(c(rep("varg", nrow(GlaucomaM)),
+         rep("vari", nrow(GlaucomaM)))))
R> panelf <- function(x, y) {
+     panel.xyplot(x, y, pch = gdata$class)
+     panel.abline(h = 0.5, lty = 2)
+ }
R> print(xyplot(avg ~ obs | var, data = gdata, panel = panelf,
+     scales = "free", xlab = "",
+     ylab = "Estimated Class Probability Glaucoma"))
```

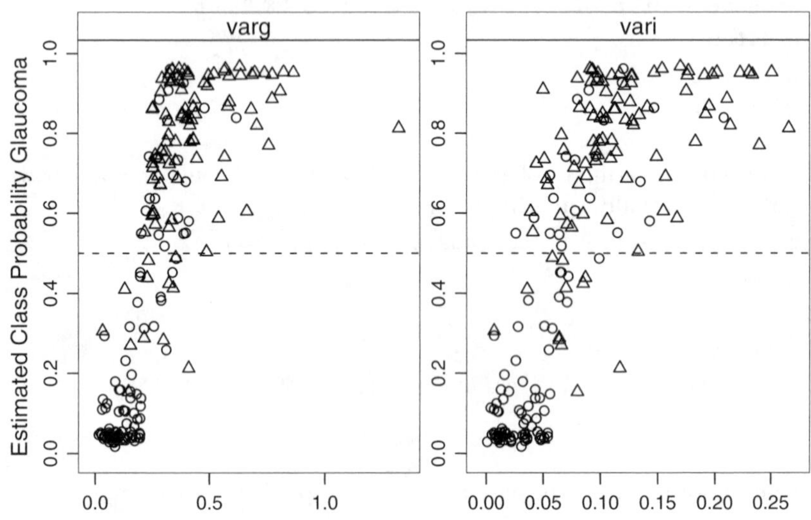


**Figure 8.4** Glaucoma data: Estimated class probabilities depending on two important variables. The 0.5 cut-off for the estimated glaucoma probability is depicted as horizontal line. Glaucomateous eyes are plotted as circles and normal eyes are triangles.

```
R> library("randomForest")
R> rf <- randomForest(Class ~ ., data = GlaucomaM)
```

and we obtain out-of-bag estimates for the prediction error via

```
R> table(predict(rf), GlaucomaM$Class)

           glaucoma normal
  glaucoma       82     12
  normal         16     86
```

```
R> plot(glaucoma_ctree)
```

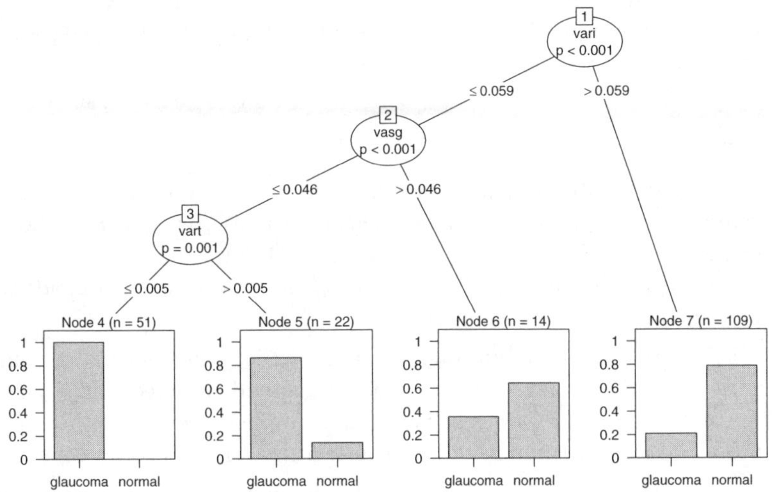


**Figure 8.5** Glaucoma data: Conditional inference tree with the distribution of glaucomateous eyes shown for each terminal leaf.

Another approach to recursive partitioning, making a connection to classical statistical test problems such as those discussed in Chapter 3, is implemented in the *party* package (Hothorn et al., 2004, 2005c). In each node of those trees, a significance test on independence between any of the covariates and the response is performed and a split is established when the $p$-value, possibly adjusted for multiple comparisons, is smaller than a pre-specified nominal level $\alpha$. This approach has the advantage that one does not need to prune back large initial trees since we have a statistically motivated stopping criterion – the $p$-value – at hand.

For the glaucoma data, such a *conditional inference tree* can be computed using the `ctree` function

```
R> library("party")
R> glaucoma_ctree <- ctree(Class ~ ., data = GlaucomaM)
```

and a graphical representation is depicted in Figure 8.5 showing both the cutpoints and the $p$-values of the associated independence tests for each node. The first split is performed using a cutpoint defined with respect to the volume of the optic nerve above some reference plane, but in the inferior part of the eye only (`vari`).

## 8.4 Summary

Recursive partitioning procedures are rather simple non-parametric tools for regression modelling. The main structures of regression relationship can be visualised in a straightforward way. However, one should bear in mind that the nature of those models is very simple and can only serve as a rough approximation to reality. When multiple simple models are averaged, powerful predictors can be constructed.

## Exercises

Ex. 8.1 Construct a classification tree for the Boston Housing data reported by Harrison and Rubinfeld (1978) which are available as *data.frame* `BostonHousing` from package *mlbench* (Leisch and Dimitriadou, 2005). Compare the predictions of the tree with the predictions obtained from `randomForest`. Which method is more accurate?

Ex. 8.2 For each possible cutpoint in `varg` of the glaucoma data, compute the test statistic of the chi-square test of independence (see Chapter 2) and plot them against the values of `varg`. Is a simple cutpoint for this variable appropriate for discriminating between healthy and glaucomateous eyes?

Ex. 8.3 Compare the tree models fitted to the glaucoma data with a logistic regression model (see Chapter 6).

CHAPTER 9

# Survival Analysis: Glioma Treatment and Breast Cancer Survival

## 9.1 Introduction

Grana et al. (2002) report results of a non-randomised clinical trial investigating a novel radioimmunotherapy in malignant glioma patients. The overall survival, i.e., the time from the beginning of the therapy to the disease-caused death of the patient, is compared for two groups of patients. A control group underwent the standard therapy and another group of patients was treated with radioimmunotherapy in addition. The data, extracted from Tables 1 and 2 in Grana et al. (2002), are given in Table 9.1. The main interest is to investigate whether the patients treated with the novel radioimmunothery survive for a longer time, compared to the patients in the control group.

**Table 9.1**: `glioma` data. Patients suffering from two types of glioma treated with the standard therapy or a novel radioimmunotherapy (`RIT`).

| `age` | `sex` | `histology` | `group` | `event` | `time` |
|---:|---:|---:|---:|---:|---:|
| 41 | Female | Grade3 | RIT | TRUE | 53 |
| 45 | Female | Grade3 | RIT | FALSE | 28 |
| 48 | Male | Grade3 | RIT | FALSE | 69 |
| 54 | Male | Grade3 | RIT | FALSE | 58 |
| 40 | Female | Grade3 | RIT | FALSE | 54 |
| 31 | Male | Grade3 | RIT | TRUE | 25 |
| 53 | Male | Grade3 | RIT | FALSE | 51 |
| 49 | Male | Grade3 | RIT | FALSE | 61 |
| 36 | Male | Grade3 | RIT | FALSE | 57 |
| 52 | Male | Grade3 | RIT | FALSE | 57 |
| 57 | Male | Grade3 | RIT | FALSE | 50 |
| 55 | Female | GBM | RIT | FALSE | 43 |
| 70 | Male | GBM | RIT | TRUE | 20 |
| 39 | Female | GBM | RIT | TRUE | 14 |
| 40 | Female | GBM | RIT | FALSE | 36 |
| 47 | Female | GBM | RIT | FALSE | 59 |
| 58 | Male | GBM | RIT | TRUE | 31 |

**Table 9.1**: `glioma` data (continued).

| age | sex | histology | group | event | time |
|---|---|---|---|---|---|
| 40 | Female | GBM | RIT | TRUE | 14 |
| 36 | Male | GBM | RIT | TRUE | 36 |
| 27 | Male | Grade3 | Control | TRUE | 34 |
| 32 | Male | Grade3 | Control | TRUE | 32 |
| 53 | Female | Grade3 | Control | TRUE | 9 |
| 46 | Male | Grade3 | Control | TRUE | 19 |
| 33 | Female | Grade3 | Control | FALSE | 50 |
| 19 | Female | Grade3 | Control | FALSE | 48 |
| 32 | Female | GBM | Control | TRUE | 8 |
| 70 | Male | GBM | Control | TRUE | 8 |
| 72 | Male | GBM | Control | TRUE | 11 |
| 46 | Male | GBM | Control | TRUE | 12 |
| 44 | Male | GBM | Control | TRUE | 15 |
| 83 | Female | GBM | Control | TRUE | 5 |
| 57 | Female | GBM | Control | TRUE | 8 |
| 71 | Female | GBM | Control | TRUE | 8 |
| 61 | Male | GBM | Control | TRUE | 6 |
| 65 | Male | GBM | Control | TRUE | 14 |
| 50 | Male | GBM | Control | TRUE | 13 |
| 42 | Female | GBM | Control | TRUE | 25 |

*Source*: From Grana, C., et. al., *Br. J. Cancer*, 86, 207–212, 2002. With permission.

The effects of hormonal treatment with Tamoxifen in women suffering from node-positive breast cancer were investigated in a randomised clinical trial as reported by Schumacher et al. (1994). Data from randomised patients from this trial and additional non-randomised patients (from the German Breast Cancer Study Group 2, GBSG2) are analysed by Sauerbrei and Royston (1999). Complete data of seven prognostic factors of 686 women are used in Sauerbrei and Royston (1999) for prognostic modelling. Observed hypothetical prognostic factors are age, menopausal status, tumor size, tumor grade, number of positive lymph nodes, progesterone receptor, estrogen receptor and the information of whether or not a hormonal therapy was applied. We are interested in an assessment of the impact of the covariates on the survival time of the patients. A subset of the patient data are shown in Table 9.2.

## 9.2 Survival Analysis

In many medical studies, the main outcome variable is the time to the occurrence of a particular event. In a randomised controlled trial of cancer, for example, surgery, radiation and chemotherapy might be compared with re-

**Table 9.2**: `GBSG2` data (package *ipred*). Randomised clinical trial data from patients suffering from node-positive breast cancer. Only the data of the first 20 patients are shown here.

| horTh | age | menostat | tsize | tgrade | pnodes | progrec | estrec | time | cens |
|---|---|---|---|---|---|---|---|---|---|
| no | 70 | Post | 21 | II | 3 | 48 | 66 | 1814 | 1 |
| yes | 56 | Post | 12 | II | 7 | 61 | 77 | 2018 | 1 |
| yes | 58 | Post | 35 | II | 9 | 52 | 271 | 712 | 1 |
| yes | 59 | Post | 17 | II | 4 | 60 | 29 | 1807 | 1 |
| no | 73 | Post | 35 | II | 1 | 26 | 65 | 772 | 1 |
| no | 32 | Pre | 57 | III | 24 | 0 | 13 | 448 | 1 |
| yes | 59 | Post | 8 | II | 2 | 181 | 0 | 2172 | 0 |
| no | 65 | Post | 16 | II | 1 | 192 | 25 | 2161 | 0 |
| no | 80 | Post | 39 | II | 30 | 0 | 59 | 471 | 1 |
| no | 66 | Post | 18 | II | 7 | 0 | 3 | 2014 | 0 |
| yes | 68 | Post | 40 | II | 9 | 16 | 20 | 577 | 1 |
| yes | 71 | Post | 21 | II | 9 | 0 | 0 | 184 | 1 |
| yes | 59 | Post | 58 | II | 1 | 154 | 101 | 1840 | 0 |
| no | 50 | Post | 27 | III | 1 | 16 | 12 | 1842 | 0 |
| yes | 70 | Post | 22 | II | 3 | 113 | 139 | 1821 | 0 |
| no | 54 | Post | 30 | II | 1 | 135 | 6 | 1371 | 1 |
| no | 39 | Pre | 35 | I | 4 | 79 | 28 | 707 | 1 |
| yes | 66 | Post | 23 | II | 1 | 112 | 225 | 1743 | 0 |
| yes | 69 | Post | 25 | I | 1 | 131 | 196 | 1781 | 0 |
| no | 55 | Post | 65 | I | 4 | 312 | 76 | 865 | 1 |
| ⋮ | ⋮ | ⋮ | ⋮ | ⋮ | ⋮ | ⋮ | ⋮ | ⋮ | ⋮ |

*Source*: From Sauerbrei, W. and Royston, P., *J. Roy. Stat. Soc. A*, 162, 71–94, 1999. With permission.

spect to time from randomisation and the start of therapy until death. In this case, the event of interest is the death of a patient, but in other situations, it might be remission from a disease, relief from symptoms or the recurrence of a particular condition. Other censored response variables are the time to credit failure in financial applications or the time a roboter needs to successfully perform a certain task in engineering. Such observations are generally referred to by the generic term *survival data* even when the endpoint or event being considered is not death but something else. Such data generally require special techniques for analysis for two main reasons:

1. Survival data are generally not symmetrically distributed – they will often appear positively skewed, with a few people surviving a very long time compared with the majority; so assuming a normal distribution will not be reasonable.
2. At the completion of the study, some patients may not have reached the endpoint of interest (death, relapse, etc.). Consequently, the exact survival times are not known. All that is known is that the survival times are greater than the amount of time the individual has been in the study. The survival times of these individuals are said to be *censored* (precisely, they are right-censored).

Of central importance in the analysis of survival time data are two functions used to describe their distribution, namely the *survival* (or *survivor*) *function* and the *hazard function.*

### *9.2.1 The Survivor Function*

The survivor function, $S(t)$, is defined as the probability that the survival time, $T$, is greater than or equal to some time $t$, i.e.,

$$S(t) = \mathsf{P}(T \geq t)$$

A plot of an estimate $\hat{S}(t)$ of $S(t)$ against the time $t$ is often a useful way of describing the survival experience of a group of individuals. When there are no censored observations in the sample of survival times, a non-parametric survivor function can be estimated simply as

$$\hat{S}(t) = \frac{\text{number of individuals with survival times} \geq t}{n}$$

where $n$ is the total number of observations. Because this is simply a proportion, confidence intervals can be obtained for each time $t$ by using the variance estimate

$$\hat{S}(t)(1 - \hat{S}(t))/n.$$

The simple method used to estimate the survivor function when there are no censored observations cannot now be used for survival times when censored observations are present. In the presence of censoring, the survivor function is typically estimated using the *Kaplan-Meier* estimator (Kaplan and Meier,

1958). This involves first ordering the survival times from the smallest to the largest such that $t_{(1)} \leq t_{(2)} \leq \dots \leq t_{(n)}$, where $t_{(j)}$ is the $j$th largest unique survival time. The Kaplan-Meier estimate of the survival function is obtained as

$$\hat{S}(t) = \prod_{j:t_{(j)} \leq t} \left(1 - \frac{d_j}{r_j}\right)$$

where $r_j$ is the number of individuals at risk just before $t_{(j)}$ (including those censored at $t_{(j)}$), and $d_j$ is the number of individuals who experience the event of interest (death, etc.) at time $t_{(j)}$. So, for example, the survivor function at the second death time, $t_{(2)}$ is equal to the estimated probability of not dying at time $t_{(2)}$, conditional on the individual being still at risk at time $t_{(2)}$. The estimated variance of the Kaplan-Meier estimate of the survivor function is found from

$$\mathsf{Var}(\hat{S}(t)) = \left(\hat{S}(t)\right)^2 \sum_{j:t_{(j)} \leq t} \frac{d_j}{r_j(r_j - d_j)}.$$

A formal test of the equality of the survival curves for the two groups can be made using the *log-rank test*. First, the expected number of deaths is computed for each unique death time, or *failure time* in the data set, assuming that the chances of dying, given that subjects are at risk, are the same for both groups. The total number of expected deaths is then computed for each group by adding the expected number of deaths for each failure time. The test then compares the observed number of deaths in each group with the expected number of deaths using a chi-squared test. Full details and formulae are given in Therneau and Grambsch (2000) or Everitt and Rabe-Hesketh (2001), for example.

### *9.2.2 The Hazard Function*

In the analysis of survival data it is often of interest to assess which periods have high or low chances of death (or whatever the event of interest may be), among those still active at the time. A suitable approach to characterise such risks is the hazard function, $h(t)$, defined as the probability that an individual experiences the event in a small time interval, $s$, given that the individual has survived up to the beginning of the interval, when the size of the time interval approaches zero; mathematically this is written as

$$h(t) = \lim_{s \to 0} \mathsf{P}(t \leq T \leq t + s | T \geq t)$$

where $T$ is the individual's survival time. The conditioning feature of this definition is very important. For example, the probability of dying at age 100 is very small because most people die before that age; in contrast, the probability of a person dying at age 100 who has reached that age is much greater.

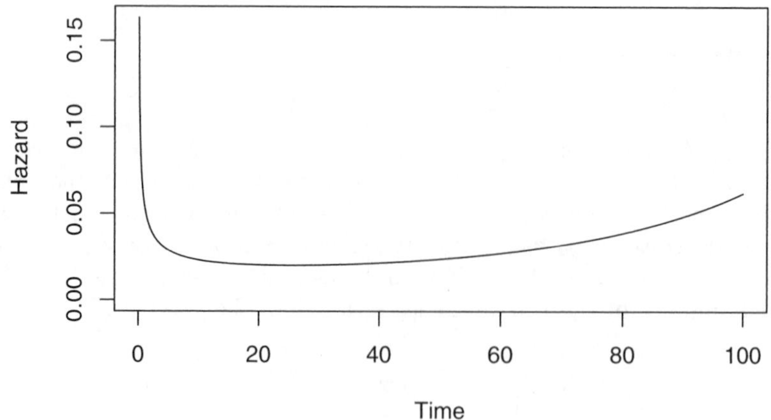


**Figure 9.1** 'Bath tub' shape of a hazard function.

The hazard function and survivor function are related by the formula

$$S(t) = \exp(-H(t))$$

where $H(t)$ is known as the *integrated hazard* or *cumulative hazard*, and is defined as follows:

$$H(t) = \int_0^t h(u)du,$$

details of how this relationship arises are given in Everitt and Pickles (2000).

In practice the hazard function may increase, decrease, remain constant or have a more complex shape. The hazard function for death in human beings, for example, has the 'bath tub' shape shown in Figure 9.1. It is relatively high immediately after birth, declines rapidly in the early years and then remains approximately constant before beginning to rise again during late middle age.

The hazard function can be estimated as the proportion of individuals experiencing the event of interest in an interval per unit time, given that they have survived to the beginning of the interval, that is

$$\hat{h}(t) = \frac{d_j}{n_j(t_{(j+1)} - t_{(j)})}.$$

The sampling variation in the estimate of the hazard function within each interval is usually considerable and so it is rarely plotted directly. Instead the integrated hazard is used. Everitt and Rabe-Hesketh (2001) show that this can be estimated as follows:

$$\hat{H}(t) = \sum_j \frac{d_j}{n_j}.$$

### 9.2.3 *Cox's Regression*

When the response variable of interest is a possibly censored survival time, we need special regression techniques for modelling the relationship of the response to explanatory variables of interest. A number of procedures are available but the most widely used by some margin is that known as *Cox's proportional hazards model*, or *Cox's regression* for short. Introduced by Sir David Cox in 1972 (see Cox, 1972), the method has become one of the most commonly used in medical statistics and the original paper one of the most heavily cited.

The main vehicle for modelling in this case is the hazard function rather than the survivor function, since it does not involve the cumulative history of events. But modelling the hazard function directly as a linear function of explanatory variables is not appropriate since $h(t)$ is restricted to being positive. A more suitable model might be

$$\log(h(t)) = \beta_0 + \beta_1 x_1 + \ldots + \beta_q x_q. \tag{9.1}$$

But this would only be suitable for a hazard function that is constant over time; this is very restrictive since hazards that increase or decrease with time, or have some more complex form are far more likely to occur in practice. In general it may be difficult to find the appropriate explicit function of time to include in (9.1). The problem is overcome in the proportional hazards model proposed by Cox (1972) by allowing the form of dependence of $h(t)$ on $t$ to remain unspecified, so that

$$\log(h(t)) = \log(h_0(t)) + \beta_1 x_1 + \ldots + \beta_q x_q$$

where $h_0(t)$ is known as the *baseline hazard function*, being the hazard function for individuals with all explanatory variables equal to zero. The model can be rewritten as

$$h(t) = h_0(t) \exp(\beta_1 x_1 + \ldots + \beta_q x_q).$$

Written in this way we see that the model forces the hazard ratio between two individuals to be constant over time since

$$\frac{h(t|\mathbf{x}_1)}{h(t|\mathbf{x}_2)} = \frac{\exp(\beta^\top \mathbf{x}_1)}{\exp(\beta^\top \mathbf{x}_2)}$$

where $\mathbf{x}_1$ and $\mathbf{x}_2$ are vectors of covariate values for two individuals. In other words, if an individual has a risk of death at some initial time point that is twice as high as another individual, then at all later times, the risk of death remains twice as high. Hence the term proportional hazards.

In the Cox model, the baseline hazard describes the common shape of the survival time distribution for all individuals, while the *relative risk function*, $\exp(\beta^\top \mathbf{x})$, gives the level of each individual's hazard. The interpretation of the parameter $\beta_j$ is that $\exp(\beta_j)$ gives the relative risk change associated with an increase of one unit in covariate $x_j$, all other explanatory variables remaining constant.

The parameters in a Cox model can be estimated by maximising what is known as a *partial likelihood.* Details are given in Kalbfleisch and Prentice (1980). The partial likelihood is derived by assuming continuous survival times. In reality, however, survival times are measured in discrete units and there are often ties. There are three common methods for dealing with ties which are described briefly in Everitt and Rabe-Hesketh (2001).

## 9.3 Analysis Using R

### *9.3.1 Glioma Radioimmunotherapy*

The survival times for patients from the control group and the group treated with the novel therapy can be compared graphically by plotting the Kaplan-Meier estimates of the survival times. Here, we plot the Kaplan-Meier estimates stratified for patients suffering from grade III glioma and glioblastoma (GBM, grade IV) separately, the results are given in Figure 9.2. The Kaplan-Meier estimates are computed by the `survfit` function from package *survival* (Therneau and Lumley, 2005) which takes a model *formula* of the form

```
Surv(time, event) ~ group
```

where `time` are the survival times, `event` is a logical variable being `TRUE` when the event of interest, death for example, has been observed and `FALSE` when in case of censoring. The right hand side variable `group` is a grouping factor.

Figure 9.2 leads to the impression that patients treated with the novel radioimmunotherapy survive longer, regardless of the tumor type. In order to assess if this informal finding is reliable, we may perform a log-rank test via

```
R> survdiff(Surv(time, event) ~ group, data = g3)

Call:
survdiff(formula = Surv(time, event) ~ group, data = g3)

               N Observed Expected (O-E)^2/E (O-E)^2/V
group=Control  6        4     1.49      4.23      6.06
group=RIT     11        2     4.51      1.40      6.06

 Chisq= 6.1  on 1 degrees of freedom, p= 0.0138
```

which indicates that the survival times are indeed different in both groups. However, the number of patients is rather limited and so it might be dangerous to rely on asymptotic tests. As shown in Chapter 3, conditioning on the data and computing the distribution of the test statistics without additional assumptions is one alternative. The function `surv_test` from package

```
R> data("glioma", package = "coin")
R> library("survival")
R> layout(matrix(1:2, ncol = 2))
R> g3 <- subset(glioma, histology == "Grade3")
R> plot(survfit(Surv(time, event) ~ group, data = g3),
+      main = "Grade III Glioma", lty = c(2, 1),
+      ylab = "Probability",
+      xlab = "Survival Time in Month", legend.bty = "n",
+      legend.text = c("Control", "Treated"))
R> g4 <- subset(glioma, histology == "GBM")
R> plot(survfit(Surv(time, event) ~ group, data = g4),
+      main = "Grade IV Glioma", ylab = "Probability",
+      lty = c(2, 1), xlab = "Survival Time in Month",
+      xlim = c(0, max(glioma$time) * 1.05))
```

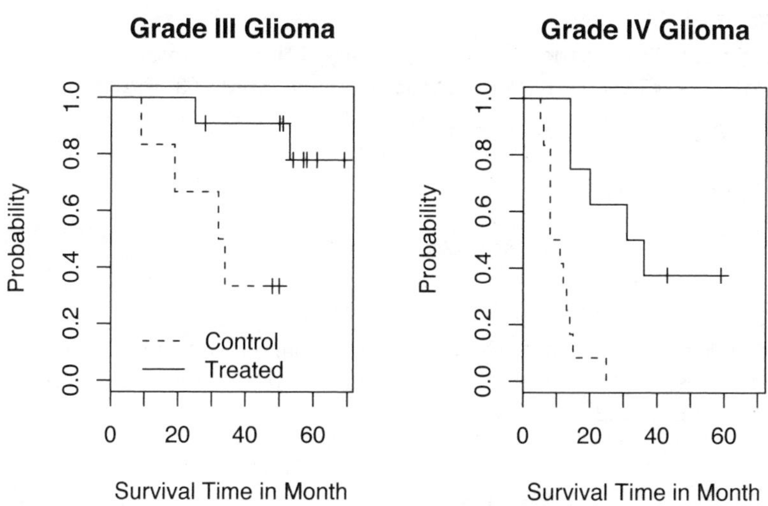


**Figure 9.2** Survival times comparing treated and control patients.

*coin* (Hothorn et al., 2005a) can be used to compute an exact conditional test answering the question whether the survival times differ for grade III patients:

```
R> library("coin")
R> surv_test(Surv(time, event) ~ group, data = g3,
+      distribution = "exact")

        Exact Logrank Test

data:  Surv(time, event) by groups Control, RIT
```

```
Z = 2.1711, p-value = 0.02877
alternative hypothesis: two.sided
```

which, in this case, confirms the above results. The same exercise can be performed for patients with grade IV glioma

```
R> surv_test(Surv(time, event) ~ group, data = g4,
+     distribution = "exact")

        Exact Logrank Test

data:  Surv(time, event) by groups Control, RIT
Z = 3.2215, p-value = 0.0001588
alternative hypothesis: two.sided
```

which shows a difference as well. However, it might be more appropriate to answer the question whether the novel therapy is superior for both groups of tumors simultaneously. This can be implemented by *stratifying*, or *blocking*, with respect tumor grading:

```
R> surv_test(Surv(time, event) ~ group | histology,
+     data = glioma, distribution = approximate(B = 10000))

        Approximative Logrank Test

data:  Surv(time, event) by
        groups Control, RIT
        stratified by histology
Z = 3.6704, p-value = 1e-04
alternative hypothesis: two.sided
```

Here, we need to approximate the exact conditional distribution since the exact distribution is hard to compute. The result supports the initial impression implied by Figure 9.2

### 9.3.2 Breast Cancer Survival

Before fitting a Cox model to the GBSG2 data, we again derive a Kaplan-Meier estimate of the survival function of the data, here stratified with respect to whether a patient received a hormonal therapy or not (see Figure 9.3).

Fitting a Cox model follows roughly the same rules are shown for linear models in Chapters 4, 5 or 6 with the exception that the response variable is again coded as a *Surv* object. For the GBSG2 data, the model is fitted via

```
R> GBSG2_coxph <- coxph(Surv(time, cens) ~ ., data = GBSG2)
```

and the results as given by the summary method are given in Figure 9.4. Since we are especially interested in the relative risk for patients who underwent a hormonal therapy, we can compute an estimate of the relative risk and a corresponding confidence interval via

```
R> ci <- confint(GBSG2_coxph)
R> exp(cbind(coef(GBSG2_coxph), ci))["horThyes", ]
```

```
R> data("GBSG2", package = "ipred")
R> plot(survfit(Surv(time, cens) ~ horTh, data = GBSG2),
+      lty = 1:2, mark.time = FALSE, ylab = "Probability",
+      xlab = "Survival Time in Days")
R> legend(250, 0.2, legend = c("yes", "no"), lty = c(2,
+      1), title = "Hormonal Therapy", bty = "n")
```

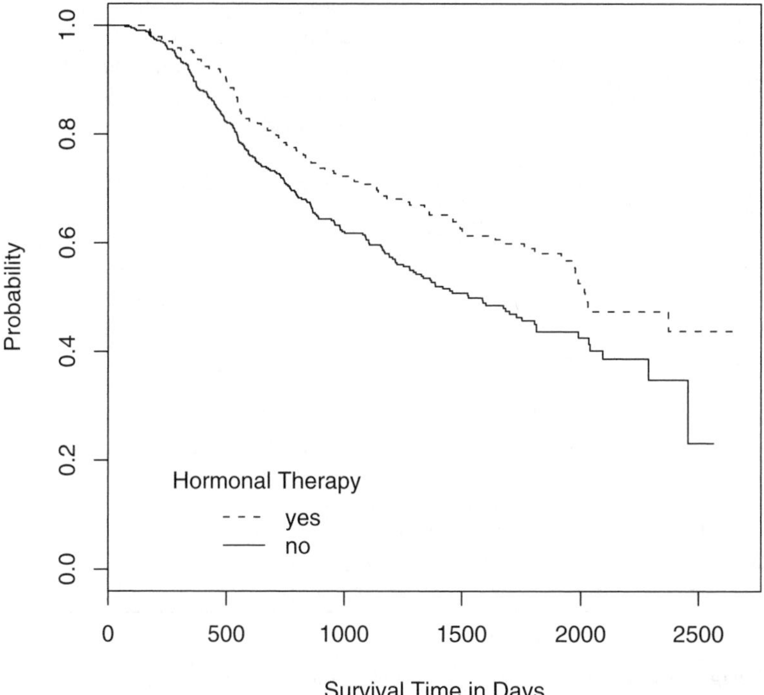


**Figure 9.3** Kaplan-Meier estimates for breast cancer patients who either received a hormonal therapy or not.

```
              2.5 %     97.5 %
0.7073155 0.5492178 0.9109233
```

This result implies that patients treated with a hormonal therapy had a lower risk and thus survived longer compared to women who were not treated this way.

Model checking and model selection for proportional hazards models are complicated by the fact that easy to use residuals, such as those discussed in

```
R> summary(GBSG2_coxph)

Call:
coxph(formula = Surv(time, cens) ~ ., data = GBSG2)

  n= 686
                  coef exp(coef) se(coef)      z       p
horThyes     -0.346278     0.707 0.129075 -2.683 7.3e-03
age          -0.009459     0.991 0.009301 -1.017 3.1e-01
menostatPost  0.258445     1.295 0.183476  1.409 1.6e-01
tsize         0.007796     1.008 0.003939  1.979 4.8e-02
tgrade.L      0.551299     1.736 0.189844  2.904 3.7e-03
tgrade.Q     -0.201091     0.818 0.121965 -1.649 9.9e-02
pnodes        0.048789     1.050 0.007447  6.551 5.7e-11
progrec      -0.002217     0.998 0.000574 -3.866 1.1e-04
estrec        0.000197     1.000 0.000450  0.438 6.6e-01

             exp(coef) exp(-coef) lower .95 upper .95
horThyes         0.707      1.414     0.549     0.911
age              0.991      1.010     0.973     1.009
menostatPost     1.295      0.772     0.904     1.855
tsize            1.008      0.992     1.000     1.016
tgrade.L         1.736      0.576     1.196     2.518
tgrade.Q         0.818      1.223     0.644     1.039
pnodes           1.050      0.952     1.035     1.065
progrec          0.998      1.002     0.997     0.999
estrec           1.000      1.000     0.999     1.001

Rsquare= 0.142   (max possible= 0.995 )
Likelihood ratio test= 105  on 9 df,   p=0
Wald test            = 115  on 9 df,   p=0
Score (logrank) test = 121  on 9 df,   p=0
```

**Figure 9.4** R output of the `summary` method for `GBSG2_coxph`.

Chapter 5 for linear regression model are not available, but several possibilities do exist. A check of the proportional hazards assumption can be done by looking at the parameter estimates $\beta_1, \dots, \beta_q$ over time. We can safely assume proportional hazards when the estimates don't vary much over time. The null hypothesis of constant regression coefficients can be tested, both globally as well as for each covariate, by using the `cox.zph` function

```
R> GBSG2_zph <- cox.zph(GBSG2_coxph)
R> GBSG2_zph

                   rho    chisq       p
horThyes     -2.54e-02 1.96e-01 0.65778
age           9.40e-02 2.96e+00 0.08552
menostatPost -1.19e-05 3.75e-08 0.99985
```

```
R> plot(GBSG2_zph, var = "age")
```

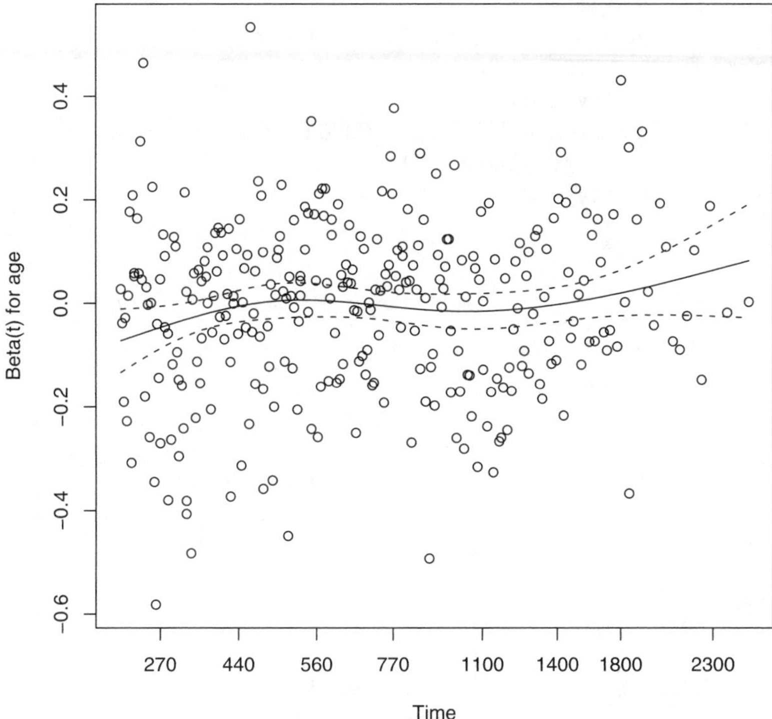


**Figure 9.5** Estimated regression coefficient for `age` depending on time for the `GBSG2` data.

```
tsize        -2.50e-02 1.88e-01 0.66436
tgrade.L     -1.30e-01 4.85e+00 0.02772
tgrade.Q      3.22e-03 3.14e-03 0.95530
pnodes        5.84e-02 5.98e-01 0.43941
progrec       5.65e-02 1.20e+00 0.27351
estrec        5.46e-02 1.03e+00 0.30967
GLOBAL              NA 2.27e+01 0.00695
```

There seems to be some evidence of time-varying effects, especially for age and tumor grading. A graphical representation of the estimated regression coefficient over time is shown in Figure 9.5. We refer to Therneau and Grambsch (2000) for a detailed theoretical description of these topics.

Martingale residuals as computed by the `residuals` method applied to `coxph` objects can be used to check the model fit. When evaluated at the

```
R> layout(matrix(1:3, ncol = 3))
R> res <- residuals(GBSG2_coxph)
R> plot(res ~ age, data = GBSG2, ylim = c(-2.5, 1.5),
+      pch = ".", ylab = "Martingale Residuals")
R> abline(h = 0, lty = 3)
R> plot(res ~ pnodes, data = GBSG2, ylim = c(-2.5,
+      1.5), pch = ".", ylab = "")
R> abline(h = 0, lty = 3)
R> plot(res ~ log(progrec), data = GBSG2, ylim = c(-2.5,
+      1.5), pch = ".", ylab = "")
R> abline(h = 0, lty = 3)
```

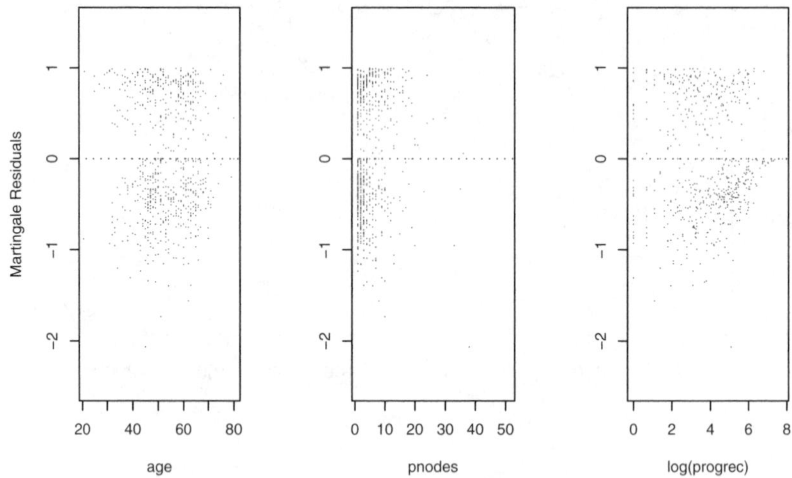


**Figure 9.6** Martingale residuals for the GBSG2 data.

true regression coefficient the expectation of the martingale residuals is zero. Thus, one way to check for systematic deviations is an inspection of scatter-plots plotting covariates against the martingale residuals. For the GBSG2 data, Figure 9.6 does not indicate severe and systematic deviations from zero.

The tree-structured regression models applied to continuous and binary responses in Chapter 8 are applicable to censored responses in survival analysis as well. Such a simple prognostic model with only a few terminal nodes might be helpful for relating the risk to certain subgroups of patients. Both rpart and the ctree function from package *party* can be applied to the GBSG2 data, where the conditional trees of the latter selects cutpoints based on log-rank statistics;

```
R> GBSG2_ctree <- ctree(Surv(time, cens) ~ ., data = GBSG2)
```

```
R> plot(GBSG2_ctree)
```

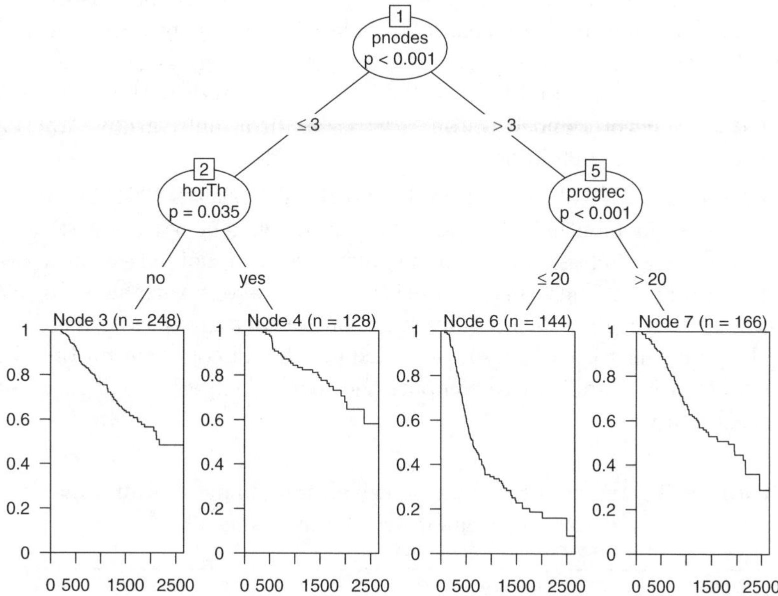


**Figure 9.7** GBSG2 data: Conditonal inference tree with the survival function, estimated by Kaplan-Meier, shown for every subgroup of patients identified by the tree.

and the `plot` method applied to this tree produces the graphical representation in Figure 9.7. The number of positive lymph nodes (`pnodes`) is the most important variable in the tree, this corresponds to the $p$-value associated with this variable in Cox's regression, see Figure 9.4. Women with not more than three positive lymph nodes who have undergone a hormonal therapy seem to have the best prognosis whereas a large number of positive lymph nodes and a small value of the progesterone receptor indicates a bad prognosis.

## 9.4 Summary

The analysis of life-time data is complicated by the fact that the time to some event is not observable for all observations due to censoring. Survival times are analysed by some estimates of the survival function, for example by a non-parametric Kaplan-Meier estimate or by semi-parametric proportional hazards regression models.

## Exercises

Ex. 9.1 Sauerbrei and Royston (1999) analyse the `GBSG2` data using multivariable fractional polynomials, a flexibilisation for many linear regression models including Cox's model. In R, this methodology is available by the *mfp* package (Ambler and Benner, 2005). Try to reproduce the analysis presented by Sauerbrei and Royston (1999), i.e., fit a multivariable fractional polynomial to the `GBSG2` data!

Ex. 9.2 The data in Table 9.3 (Everitt and Rabe-Hesketh, 2001) are the survival times (in months) after mastectomy of women with breast cancer. The cancers are classified as having metastised or not based on a histochemical marker. Censoring is indicated by the `event` variable being `TRUE` in case of death. Plot the survivor functions of each group, estimated using the Kaplan-Meier estimate, on the same graph and comment on the differences. Use a log-rank test to compare the survival experience of each group more formally.

**Table 9.3**: `mastectomy` data. Survival times in months after mastectomy of women with breast cancer.

| time | event | metastized | time | event | metastized |
|---|---|---|---|---|---|
| 23 | TRUE | no | 40 | TRUE | yes |
| 47 | TRUE | no | 41 | TRUE | yes |
| 69 | TRUE | no | 48 | TRUE | yes |
| 70 | FALSE | no | 50 | TRUE | yes |
| 100 | FALSE | no | 59 | TRUE | yes |
| 101 | FALSE | no | 61 | TRUE | yes |
| 148 | TRUE | no | 68 | TRUE | yes |
| 181 | TRUE | no | 71 | TRUE | yes |
| 198 | FALSE | no | 76 | FALSE | yes |
| 208 | FALSE | no | 105 | FALSE | yes |
| 212 | FALSE | no | 107 | FALSE | yes |
| 224 | FALSE | no | 109 | FALSE | yes |
| 5 | TRUE | yes | 113 | TRUE | yes |
| 8 | TRUE | yes | 116 | FALSE | yes |
| 10 | TRUE | yes | 118 | TRUE | yes |
| 13 | TRUE | yes | 143 | TRUE | yes |
| 18 | TRUE | yes | 145 | FALSE | yes |
| 24 | TRUE | yes | 162 | FALSE | yes |
| 26 | TRUE | yes | 188 | FALSE | yes |
| 26 | TRUE | yes | 212 | FALSE | yes |
| 31 | TRUE | yes | 217 | FALSE | yes |
| 35 | TRUE | yes | 225 | FALSE | yes |

CHAPTER 10

# Analysing Longitudinal Data I: Computerised Delivery of Cognitive Behavioural Therapy–Beat the Blues

## 10.1 Introduction

Depression is a major public health problem across the world. Antidepressants are the front line treatment, but many patients either do not respond to them, or do not like taking them. The main alternative is psychotherapy, and the modern 'talking treatments' such as *cognitive behavioural therapy* (CBT) have been shown to be as effective as drugs, and probably more so when it comes to relapse. But there is a problem, namely availability–there are simply nothing like enough skilled therapists to meet the demand, and little prospect at all of this situation changing.

A number of alternative modes of delivery of CBT have been explored, including interactive systems making use of the new computer technologies. The principles of CBT lend themselves reasonably well to computerisation, and, perhaps surprisingly, patients adapt well to this procedure, and do not seem to miss the physical presence of the therapist as much as one might expect. The data to be used in this chapter arise from a clinical trial of an interactive, multimedia program known as 'Beat the Blues' designed to deliver cognitive behavioural therapy to depressed patients via a computer terminal. Full details are given in Proudfoot et al. (2003), but in essence Beat the Blues is an interactive program using multimedia techniques, in particular video vignettes. The computer based intervention consists of nine sessions, followed by eight therapy sessions, each lasting about 50 minutes. Nurses are used to explain how the program works, but are instructed to spend no more than 5 minutes with each patient at the start of each session, and are there simply to assist with the technology. In a randomised controlled trial of the program, patients with depression recruited in primary care were randomised to either the Beating the Blues program, or to 'Treatment as Usual' (TAU). Patients randomised to Beat the Blues also received pharmacology and/or general practice (GP) support and practical/social help, offered as part of treatment as usual, with the exception of any face-to-face counselling or psychological intervention. Patients allocated to TAU received whatever treatment their GP prescribed. The latter included, besides any medication, discussion of problems with GP, provision of practical/social help, referral to a counsellor, referral to a practice nurse, referral to mental health professionals (psychologist, psychiatrist, community psychiatric nurse, counsellor), or further physical examination.

A number of outcome measures were used in the trial, but here we concentrate on the *Beck Depression Inventory II* (BDI, Beck et al., 1996). Measurements on this variable were made on the following five occasions:

- Prior to treatment,
- Two months after treatment began and
- At one, three and six months follow-up, i.e., at three, five and eight months after treatment.

**Table 10.1**: `BtheB` data. Data of a randomised trial evaluating the effects of Beat the Blues.

| drug | length | treatment | bdi.pre | bdi.2m | bdi.4m | bdi.6m | bdi.8m |
|---|---|---|---|---|---|---|---|
| No | >6m | TAU | 29 | 2 | 2 | NA | NA |
| Yes | >6m | BtheB | 32 | 16 | 24 | 17 | 20 |
| Yes | <6m | TAU | 25 | 20 | NA | NA | NA |
| No | >6m | BtheB | 21 | 17 | 16 | 10 | 9 |
| Yes | >6m | BtheB | 26 | 23 | NA | NA | NA |
| Yes | <6m | BtheB | 7 | 0 | 0 | 0 | 0 |
| Yes | <6m | TAU | 17 | 7 | 7 | 3 | 7 |
| No | >6m | TAU | 20 | 20 | 21 | 19 | 13 |
| Yes | <6m | BtheB | 18 | 13 | 14 | 20 | 11 |
| Yes | >6m | BtheB | 20 | 5 | 5 | 8 | 12 |
| No | >6m | TAU | 30 | 32 | 24 | 12 | 2 |
| Yes | <6m | BtheB | 49 | 35 | NA | NA | NA |
| No | >6m | TAU | 26 | 27 | 23 | NA | NA |
| Yes | >6m | TAU | 30 | 26 | 36 | 27 | 22 |
| Yes | >6m | BtheB | 23 | 13 | 13 | 12 | 23 |
| No | <6m | TAU | 16 | 13 | 3 | 2 | 0 |
| No | >6m | BtheB | 30 | 30 | 29 | NA | NA |
| No | <6m | BtheB | 13 | 8 | 8 | 7 | 6 |
| No | >6m | TAU | 37 | 30 | 33 | 31 | 22 |
| Yes | <6m | BtheB | 35 | 12 | 10 | 8 | 10 |
| No | >6m | BtheB | 21 | 6 | NA | NA | NA |
| No | <6m | TAU | 26 | 17 | 17 | 20 | 12 |
| No | >6m | TAU | 29 | 22 | 10 | NA | NA |
| No | >6m | TAU | 20 | 21 | NA | NA | NA |
| No | >6m | TAU | 33 | 23 | NA | NA | NA |
| No | >6m | BtheB | 19 | 12 | 13 | NA | NA |
| Yes | <6m | TAU | 12 | 15 | NA | NA | NA |
| Yes | >6m | TAU | 47 | 36 | 49 | 34 | NA |
| Yes | >6m | BtheB | 36 | 6 | 0 | 0 | 2 |
| No | <6m | BtheB | 10 | 8 | 6 | 3 | 3 |

**Table 10.1**: `BtheB` data (continued).

| drug | length | treatment | bdi.pre | bdi.2m | bdi.4m | bdi.6m | bdi.8m |
|---|---|---|---|---|---|---|---|
| No | <6m | TAU | 27 | 7 | 15 | 16 | 0 |
| No | <6m | BtheB | 18 | 10 | 10 | 6 | 8 |
| Yes | <6m | BtheB | 11 | 8 | 3 | 2 | 15 |
| Yes | <6m | BtheB | 6 | 7 | NA | NA | NA |
| Yes | >6m | BtheB | 44 | 24 | 20 | 29 | 14 |
| No | <6m | TAU | 38 | 38 | NA | NA | NA |
| No | <6m | TAU | 21 | 14 | 20 | 1 | 8 |
| Yes | >6m | TAU | 34 | 17 | 8 | 9 | 13 |
| Yes | <6m | BtheB | 9 | 7 | 1 | NA | NA |
| Yes | >6m | TAU | 38 | 27 | 19 | 20 | 30 |
| Yes | <6m | BtheB | 46 | 40 | NA | NA | NA |
| No | <6m | TAU | 20 | 19 | 18 | 19 | 18 |
| Yes | >6m | TAU | 17 | 29 | 2 | 0 | 0 |
| No | >6m | BtheB | 18 | 20 | NA | NA | NA |
| Yes | >6m | BtheB | 42 | 1 | 8 | 10 | 6 |
| No | <6m | BtheB | 30 | 30 | NA | NA | NA |
| Yes | <6m | BtheB | 33 | 27 | 16 | 30 | 15 |
| No | <6m | BtheB | 12 | 1 | 0 | 0 | NA |
| Yes | <6m | BtheB | 2 | 5 | NA | NA | NA |
| No | >6m | TAU | 36 | 42 | 49 | 47 | 40 |
| No | <6m | TAU | 35 | 30 | NA | NA | NA |
| No | <6m | BtheB | 23 | 20 | NA | NA | NA |
| No | >6m | TAU | 31 | 48 | 38 | 38 | 37 |
| Yes | <6m | BtheB | 8 | 5 | 7 | NA | NA |
| Yes | <6m | TAU | 23 | 21 | 26 | NA | NA |
| Yes | <6m | BtheB | 7 | 7 | 5 | 4 | 0 |
| No | <6m | TAU | 14 | 13 | 14 | NA | NA |
| No | <6m | TAU | 40 | 36 | 33 | NA | NA |
| Yes | <6m | BtheB | 23 | 30 | NA | NA | NA |
| No | >6m | BtheB | 14 | 3 | NA | NA | NA |
| No | >6m | TAU | 22 | 20 | 16 | 24 | 16 |
| No | >6m | TAU | 23 | 23 | 15 | 25 | 17 |
| No | <6m | TAU | 15 | 7 | 13 | 13 | NA |
| No | >6m | TAU | 8 | 12 | 11 | 26 | NA |
| No | >6m | BtheB | 12 | 18 | NA | NA | NA |
| No | >6m | TAU | 7 | 6 | 2 | 1 | NA |
| Yes | <6m | TAU | 17 | 9 | 3 | 1 | 0 |
| Yes | <6m | BtheB | 33 | 18 | 16 | NA | NA |
| No | <6m | TAU | 27 | 20 | NA | NA | NA |
| No | <6m | BtheB | 27 | 30 | NA | NA | NA |
| No | <6m | BtheB | 9 | 6 | 10 | 1 | 0 |
| No | >6m | BtheB | 40 | 30 | 12 | NA | NA |

**Table 10.1**: `BtheB` data (continued).

| drug | length | treatment | bdi.pre | bdi.2m | bdi.4m | bdi.6m | bdi.8m |
|---|---|---|---|---|---|---|---|
| No | >6m | TAU | 11 | 8 | 7 | NA | NA |
| No | <6m | TAU | 9 | 8 | NA | NA | NA |
| No | >6m | TAU | 14 | 22 | 21 | 24 | 19 |
| Yes | >6m | BtheB | 28 | 9 | 20 | 18 | 13 |
| No | >6m | BtheB | 15 | 9 | 13 | 14 | 10 |
| Yes | >6m | BtheB | 22 | 10 | 5 | 5 | 12 |
| No | <6m | TAU | 23 | 9 | NA | NA | NA |
| No | >6m | TAU | 21 | 22 | 24 | 23 | 22 |
| No | >6m | TAU | 27 | 31 | 28 | 22 | 14 |
| Yes | >6m | BtheB | 14 | 15 | NA | NA | NA |
| No | >6m | TAU | 10 | 13 | 12 | 8 | 20 |
| Yes | <6m | TAU | 21 | 9 | 6 | 7 | 1 |
| Yes | >6m | BtheB | 46 | 36 | 53 | NA | NA |
| No | >6m | BtheB | 36 | 14 | 7 | 15 | 15 |
| Yes | >6m | BtheB | 23 | 17 | NA | NA | NA |
| Yes | >6m | TAU | 35 | 0 | 6 | 0 | 1 |
| Yes | <6m | BtheB | 33 | 13 | 13 | 10 | 8 |
| No | <6m | BtheB | 19 | 4 | 27 | 1 | 2 |
| No | <6m | TAU | 16 | NA | NA | NA | NA |
| Yes | <6m | BtheB | 30 | 26 | 28 | NA | NA |
| Yes | <6m | BtheB | 17 | 8 | 7 | 12 | NA |
| No | >6m | BtheB | 19 | 4 | 3 | 3 | 3 |
| No | >6m | BtheB | 16 | 11 | 4 | 2 | 3 |
| Yes | >6m | BtheB | 16 | 16 | 10 | 10 | 8 |
| Yes | <6m | TAU | 28 | NA | NA | NA | NA |
| No | >6m | BtheB | 11 | 22 | 9 | 11 | 11 |
| No | <6m | TAU | 13 | 5 | 5 | 0 | 6 |
| Yes | <6m | TAU | 43 | NA | NA | NA | NA |

The resulting data from a subset of 100 patients are shown in Table 10.1. (The data are used with the kind permission of Dr. Judy Proudfoot.) In addition to assessing the effects of treatment, there is interest here in assessing the effect of taking antidepressant drugs (`drug`, yes or no) and length of the current episode of depression (`length`, less or more than six months).

## 10.2 Analysing Longitudinal Data

The distinguishing feature of a longitudinal study is that the response variable of interest and a set of explanatory variables are measured several times on each individual in the study. The main objective in such a study is to characterise change in the repeated values of the response variable and to determine the explanatory variables most associated with any change. Because several observations of the response variable are made on the same individual, it is likely that the measurements will be correlated rather than independent, even after conditioning on the explanatory variables. Consequently repeated measures data require special methods of analysis and models for such data need to include parameters linking the explanatory variables to the repeated measurements, parameters analogous to those in the usual multiple regression model (see Chapter 5), and, in addition parameters that account for the correlational structure of the repeated measurements. It is the former parameters that are generally of most interest with the latter often being regarded as *nuisance parameters*. But providing an adequate description for the correlational structure of the repeated measures is necessary to avoid misleading inferences about the parameters that *are* of real interest to the researcher.

Over the last decade methodology for the analysis of repeated measures data has been the subject of much research and development, and there are now a variety of powerful techniques available. A comprehensive account of these methods is given in Diggle et al. (2003) and Davis (2002). In this chapter we will concentrate on a single class of methods, *linear mixed effects models* and then in Chapter 11, describe *generalised estimating equations*, another class of models suitable for analysing longitudinal data.

## 10.3 Linear Mixed Effects Models for Repeated Measures Data

Linear mixed effects models for repeated measures data formalise the sensible idea that an individual's pattern of responses is likely to depend on many characteristics of that individual, including some that are unobserved. These unobserved variables are then included in the model as random variables, i.e., random effects. The essential feature of such models is that correlation amongst the repeated measurements on the same unit arises from shared, unobserved variables. Conditional on the values of the random effects, the repeated measurements are assumed to be independent, the so-called *local independence* assumption.

Two commonly used linear mixed effect models, the *random intercept* and the *random intercept and slope* models, will now be described in more detail.

Let $y_{ij}$ represent the observation made at time $t_j$ on individual $i$. A possible model for the observation $y_{ij}$ might be

$$y_{ij} = \beta_0 + \beta_1 t_j + u_i + \varepsilon_{ij}. \tag{10.1}$$

Here the total residual that would be present in the usual linear regression model has been partitioned into a subject-specific random component $u_i$ which is constant over time plus a residual $\varepsilon_{ij}$ which varies randomly over time. The $u_i$ are assumed to be normally distributed with zero mean and variance $\sigma_u^2$. Similarly the residuals $\varepsilon_{ij}$ are assumed normally distributed with zero mean and variance $\sigma^2$. The $u_i$ and $\varepsilon_{ij}$ are assumed to be independent of each other and of the time $t_j$. The model in (10.1) is known as a *random intercept model*, the $u_i$ being the random intercepts. The repeated measurements for an individual vary about that individual's own regression line which can differ in intercept but not in slope from the regression lines of other individuals. The random effects model possible heterogeneity in the intercepts of the individuals whereas time has a fixed effect, $\beta_1$.

The random intercept model implies that the total variance of each repeated measurement is $\mathsf{Var}(y_{ij}) = \mathsf{Var}(u_i + \varepsilon_{ij}) = \sigma_u^2 + \sigma^2$. Due to this decomposition of the total residual variance into a between-subject component, $\sigma_u^2$, and a within-subject component, $\sigma^2$, the model is sometimes referred to as a *variance component model.*

The covariance between the total residuals at two time points $j$ and $k$ in the same individual is $\mathsf{Cov}(u_i + \varepsilon_{ij}, u_i + \varepsilon_{ik}) = \sigma_u^2$. Note that these covariances are induced by the shared random intercept; for individuals with $u_i > 0$, the total residuals will tend to be greater than the mean, for individuals with $u_i < 0$ they will tend to be less than the mean. It follows from the two relations above that the residual correlations are given by

$$\mathsf{Cor}(u_i + \varepsilon_{ij}, u_i + \varepsilon_{ik}) = \frac{\sigma_u^2}{\sigma_u^2 + \sigma^2}.$$

This is an *intra-class correlation* interpreted as the proportion of the total residual variance that is due to residual variability between subjects. A random intercept model constrains the variance of each repeated measure to be the same and the covariance between any pair of measurements to be equal. This is usually called the *compound symmetry* structure. These constraints are often not realistic for repeated measures data. For example, for longitudinal data it is more common for measures taken closer to each other in time to be more highly correlated than those taken further apart. In addition the variances of the later repeated measures are often greater than those taken earlier. Consequently for many such data sets the random intercept model will not do justice to the observed pattern of covariances between the repeated measures. A model that allows a more realistic structure for the covariances is one that allows heterogeneity in both slopes and intercepts, the *random slope and intercept model.*

In this model there are two types of random effects, the first modelling heterogeneity in intercepts, $u_i$, and the second modelling heterogeneity in slopes, $v_i$. Explicitly the model is

$$y_{ij} = \beta_0 + \beta_1 t_j + u_i + v_i t_j + \varepsilon_{ij} \tag{10.2}$$

where the parameters are not, of course, the same as in (10.1). The two random effects are assumed to have a bivariate normal distribution with zero means for both variables and variances $\sigma_u^2$ and $\sigma_v^2$ with covariance $\sigma_{uv}$. With this model the total residual is $u_i + u_i t_j + \varepsilon_{ij}$ with variance

$$\mathsf{Var}(u_i + v_i t_j + \varepsilon_{ij}) = \sigma_u^2 + 2\sigma_{uv} t_j + \sigma_v^2 t_j^2 + \sigma^2$$

which is no longer constant for different values of $t_j$. Similarly the covariance between two total residuals of the same individual

$$\text{Cov}(u_i + v_i t_j + \varepsilon_{ij}, u_i + v_i t_k + \varepsilon_{ik}) = \sigma_u^2 + \sigma_{uv}(t_j - t_k) + \sigma_v^2 t_j t_k$$

is not constrained to be the same for all pairs $t_j$ and $t_k$.

(It should also be noted that re-estimating the model after adding or subtracting a constant from $t_j$, e.g., its mean, will lead to different variance and covariance estimates, but will not affect fixed effects.)

Linear mixed-effects models can be estimated by maximum likelihood. However, this method tends to underestimate the variance components. A modified version of maximum likelihood, known as *restricted maximum likelihood* is therefore often recommended; this provides consistent estimates of the variance components. Details are given in Diggle et al. (2003) and Longford (1993). Competing linear mixed-effects models can be compared using a likelihood ratio test. If however the models have been estimated by restricted maximum likelihood this test can only be used if both models have the same set of fixed effects (see Longford, 1993).

## 10.4 Analysis Using R

Almost all statistical analyses should begin with some graphical representation of the data and here we shall construct the boxplots of each of the five repeated measures separately for each treatment group. The data are available as the data frame `BtheB` and the necessary R code is given along with Figure 10.1. The boxplots show that there is decline in BDI values in both groups with perhaps the values in the group of patients treated in the 'Beat the Blues' arm being lower at each post-randomisation visit.

We shall fit both random intercept and random intercept and slope models to the data including the baseline BDI values (`pre.bdi`), `treatment` group, `drug` and `length` as fixed effect covariates. Linear mixed effects models are fitted in R by using the `lmer` function contained in the *lme4* package (Bates and Sarkar, 2005, Pinheiro and Bates, 2000, Bates, 2005), but an essential first step is to rearrange the data from the 'wide form' in which they appear in the `BtheB` data frame into the 'long form' in which each separate repeated measurement and associated covariate values appear as a separate row in a *data.frame*. This rearrangement can be made using the following code:

```
R> data("BtheB", package = "HSAUR")
R> BtheB$subject <- factor(rownames(BtheB))
R> nobs <- nrow(BtheB)
R> BtheB_long <- reshape(BtheB, idvar = "subject",
+      varying = c("bdi.2m", "bdi.4m", "bdi.6m", "bdi.8m"),
+      direction = "long")
R> BtheB_long$time <- rep(c(2, 4, 6, 8), rep(nobs,
+      4))
```

such that the data are now in the form (here shown for the first three subjects)

```
R> subset(BtheB_long, subject %in% c("1", "2", "3"))
```

```
R> data("BtheB", package = "HSAUR")
R> layout(matrix(1:2, nrow = 1))
R> ylim <- range(BtheB[, grep("bdi", names(BtheB))],
+      na.rm = TRUE)
R> boxplot(subset(BtheB, treatment == "TAU")[, grep("bdi",
+      names(BtheB))], main = "Treated as usual", ylab = "BDI",
+      xlab = "Time (in months)", names = c(0, 2, 4,
+          6, 8), ylim = ylim)
R> boxplot(subset(BtheB, treatment == "BtheB")[, grep("bdi",
+      names(BtheB))], main = "Beat the Blues", ylab = "BDI",
+      xlab = "Time (in months)", names = c(0, 2, 4,
+          6, 8), ylim = ylim)
```

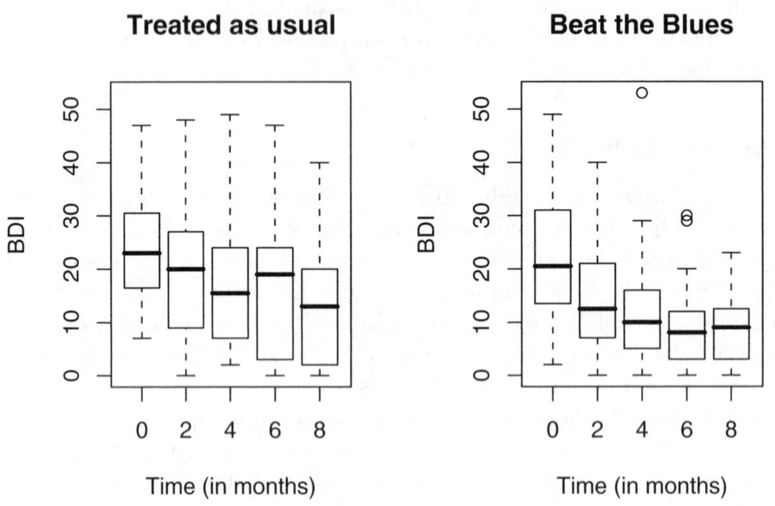


**Figure 10.1** Boxplots for the repeated measures by treatment group for the BtheB data.

| | drug | length | treatment | bdi.pre | subject | time | bdi |
|---|---|---|---|---|---|---|---|
| 1.2m | No | >6m | TAU | 29 | 1 | 2 | 2 |
| 2.2m | Yes | >6m | BtheB | 32 | 2 | 2 | 16 |
| 3.2m | Yes | <6m | TAU | 25 | 3 | 2 | 20 |
| 1.4m | No | >6m | TAU | 29 | 1 | 4 | 2 |
| 2.4m | Yes | >6m | BtheB | 32 | 2 | 4 | 24 |
| 3.4m | Yes | <6m | TAU | 25 | 3 | 4 | NA |
| 1.6m | No | >6m | TAU | 29 | 1 | 6 | NA |
| 2.6m | Yes | >6m | BtheB | 32 | 2 | 6 | 17 |
| 3.6m | Yes | <6m | TAU | 25 | 3 | 6 | NA |
| 1.8m | No | >6m | TAU | 29 | 1 | 8 | NA |

```
2.8m  Yes     >6m      BtheB        32          2     8  20
3.8m  Yes     <6m         TAU       25          3     8  NA
```

The resulting *data.frame* BtheB_long contains a number of missing values and in applying the lmer function these will be dropped. But notice it is only the missing values that are removed, *not* participants that have at least one missing value. All the available data is used in the model fitting process. The lmer function is used in a similar way to the lm function met in Chapter 5 with the addition of a random term to identify the source of the repeated measurements, here subject. We can fit the two models (10.1) and (10.2) and test which is most appropriate using

```
R> library("lme4")
R> BtheB_lmer1 <- lmer(bdi ~ bdi.pre + time + treatment +
+     drug + length + (1 | subject), data = BtheB_long,
+     method = "ML", na.action = na.omit)
R> BtheB_lmer2 <- lmer(bdi ~ bdi.pre + time + treatment +
+     drug + length + (time | subject), data = BtheB_long,
+     method = "ML", na.action = na.omit)
R> anova(BtheB_lmer1, BtheB_lmer2)
```

```
Data: BtheB_long
Models:
BtheB_lmer1: bdi ~ bdi.pre + time + treatment + drug + length
    + (1 | subject)
BtheB_lmer2: bdi ~ bdi.pre + time + treatment + drug + length
    + (time | subject)
            Df     AIC     BIC  logLik  Chisq Chi Df
BtheB_lmer1  8 1886.62 1915.70 -935.31
BtheB_lmer2 10 1889.81 1926.16 -934.90 0.8161      2
            Pr(>Chisq)
BtheB_lmer1
BtheB_lmer2      0.665
```

The log-likelihood test indicates that the simpler random intercept model is adequate for these data. More information about the fitted random intercept model can be extracted from object BtheB_lmer1 using summary by the R code in Figure 10.2. We see that the regression coefficients for time and the *Beck Depression Inventory II* values measured at baseline (bdi.pre) are highly significant, but there is no evidence that the coefficients for the other three covariates differ from zero. In particular, there is no clear evidence of a treatment effect.

We can check the assumptions of the final model fitted to the BtheB data, i.e., the normality of the random effect terms and the residuals, by first using the ranef method to *predict* the former and the residuals method to calculate the differences between the observed data values and the fitted values, and then using normal probability plots on each. How the random effects are predicted is explained briefly in Section 10.5. The necessary R code to obtain

```
R> summary(BtheB_lmer1)

Linear mixed-effects model fit by maximum likelihood
Formula: bdi ~ bdi.pre + time + treatment + drug + length + (1
    | subject)
   Data: BtheB_long
      AIC      BIC    logLik MLdeviance REMLdeviance
 1886.624 1915.702 -935.312   1870.624     1866.149
Random effects:
 Groups   Name        Variance Std.Dev.
 subject  (Intercept) 49.362   7.0258
 Residual             25.678   5.0673
# of obs: 280, groups: subject, 97

Fixed effects:
                 Estimate Std. Error  DF t value  Pr(>|t|)
(Intercept)      5.943659   2.249224 274  2.6425  0.008702 **
bdi.pre          0.638192   0.077591 274  8.2250 7.928e-15 ***
time            -0.717018   0.146055 274 -4.9092 1.573e-06 ***
treatmentBtheB  -2.373078   1.663747 274 -1.4263  0.154907
drugYes         -2.797837   1.719997 274 -1.6267  0.104960
length>6m        0.256348   1.632189 274  0.1571  0.875315
---
Signif. codes:  0 '***' 0.001 '**' 0.01 '*' 0.05 '.' 0.1 ' ' 1

Correlation of Fixed Effects:
            (Intr) bdi.pr time   trtmBB drugYs
bdi.pre     -0.678
time        -0.264  0.023
tretmntBthB -0.389  0.121  0.022
drugYes     -0.071 -0.237 -0.025 -0.323
length>6m   -0.238 -0.242 -0.043  0.002  0.158
```

**Figure 10.2** R output of the linear mixed-effects model fit for the BtheB data.

the effects, residuals and plots is shown with Figure 10.3. There appear to be no large departures from linearity in either plot.

## 10.5 Prediction of Random Effects

The random effects are not estimated as part of the model. However, having estimated the model, we can *predict* the values of the random effects. According to Bayes' Theorem, the *posterior probability* of the random effects is given by

$$\mathsf{P}(u|y, x) = f(y|u, x)g(u)$$

```
R> layout(matrix(1:2, ncol = 2))
R> qint <- ranef(BtheB_lmer1)$subject[["(Intercept)"]]
R> qres <- residuals(BtheB_lmer1)
R> qqnorm(qint, ylab = "Estimated random intercepts",
+      xlim = c(-3, 3), ylim = c(-20, 20),
+      main = "Random intercepts")
R> qqline(qint)
R> qqnorm(qres, xlim = c(-3, 3), ylim = c(-20, 20),
+      ylab = "Estimated residuals", main = "Residuals")
R> qqline(qres)
```

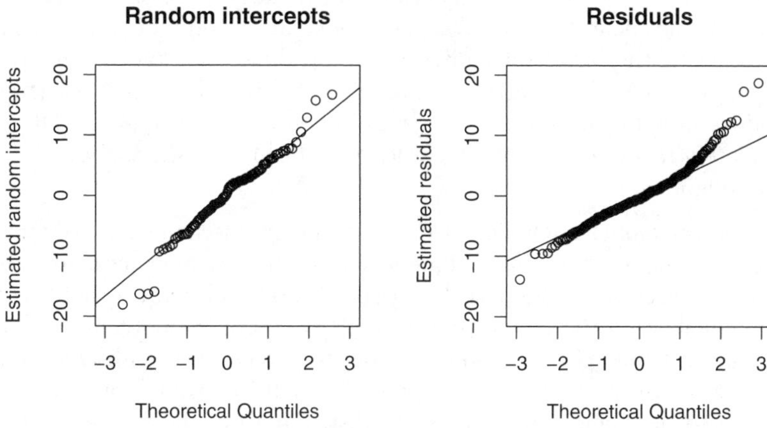


**Figure 10.3** Quantile-quantile plots of predicted random intercepts and residuals for the random intercept model `BtheB_lmer1` fitted to the `BtheB` data.

where $f(y|u,x)$ is the conditional density of the responses given the random effects and covariates (a product of normal densities) and $g(u)$ is the *prior* density of the random effects (multivariate normal). The means of this posterior distribution can be used as estimates of the random effects and are known as *empirical Bayes estimates.* The empirical Bayes estimator is also known as a shrinkage estimator because the predicted random effects are smaller in absolute value than their fixed effect counterparts. *Best linear unbiased predictions* (BLUP) are linear combinations of the responses that are unbiased estimators of the random effects and minimise the mean square error.

## 10.6 The Problem of Dropouts

We now need to consider briefly how the dropouts may affect the analyses reported above. To understand the problems that patients dropping out can cause for the analysis of data from a longitudinal trial we need to consider a classification of dropout mechanisms first introduced by Rubin (1976). The type of mechanism involved has implications for which approaches to analysis are suitable and which are not. Rubin's suggested classification involves three types of dropout mechanism:

*Dropout completely at random* (DCAR): here the probability that a patient drops out does not depend on either the observed or missing values of the response. Consequently the observed (non-missing) values effectively constitute a simple random sample of the values for all subjects. Possible examples include missing laboratory measurements because of a dropped test-tube (if it was not dropped because of the knowledge of any measurement), the accidental death of a participant in a study, or a participant moving to another area. Intermittent missing values in a longitudinal data set, whereby a patient misses a clinic visit for transitory reasons ('went shopping instead' or the like) can reasonably be assumed to be DCAR. Completely random dropout causes least problem for data analysis, but it is a strong assumption.

*Dropout at random* (DAR): The dropout at random mechanism occurs when the probability of dropping out depends on the outcome measures that have been observed in the past, but given this information is conditionally independent of all the future (unrecorded) values of the outcome variable following dropout. Here 'missingness' depends only on the observed data with the distribution of future values for a subject who drops out at a particular time being the same as the distribution of the future values of a subject who remains in at that time, if they have the same covariates and the same past history of outcome up to and including the specific time point. Murray and Findlay (1988) provide an example of this type of missing value from a study of hypertensive drugs in which the outcome measure was diastolic blood pressure. The protocol of the study specified that the participant was to be removed from the study when his/her blood pressure got too large. Here blood pressure at the time of dropout was observed before the participant dropped out, so although the dropout mechanism is not DCAR since it depends on the values of blood pressure, it *is* DAR, because dropout depends only on the observed part of the data. A further example of a DAR mechanism is provided by Heitjan (1997), and involves a study in which the response measure is body mass index (BMI). Suppose that the measure is missing because subjects who had high body mass index values at earlier visits avoided being measured at later visits out of embarrassment, regardless of whether they had gained or lost weight in the intervening period. The missing values here are DAR but *not* DCAR; consequently methods applied to the data that assumed the latter might give misleading results (see later discussion).

*Non-ignorable* (sometimes referred to as *informative*): The final type of dropout mechanism is one where the probability of dropping out depends on the unrecorded missing values – observations are likely to be missing when the outcome values that would have been observed had the patient not dropped out, are systematically higher or lower than usual (corresponding perhaps to their condition becoming worse or improving). A non-medical example is when individuals with lower income levels or very high incomes are less likely to provide their personal income in an interview. In a medical setting possible examples are a participant dropping out of a longitudinal study when his/her blood pressure became too high and this value was not observed, or when their pain become intolerable and we did not record the associated pain value. For the BDI example introduced above, if subjects were more likely to avoid being measured if they had put on extra weight since the last visit, then the data are non-ignorably missing. Dealing with data containing missing values that result from this type of dropout mechanism is difficult. The correct analyses for such data must estimate the dependence of the missingness probability on the missing values. Models and software that attempt this are available (see, for example, Diggle and Kenward, 1994) but their use is not routine and, in addition, it must be remembered that the associated parameter estimates can be unreliable.

Under what type of dropout mechanism are the mixed effects models considered in this chapter valid? The good news is that such models can be shown to give valid results under the relatively weak assumption that the dropout mechanism is DAR (see Carpenter et al., 2002). When the missing values are thought to be informative, any analysis is potentially problematical. But Diggle and Kenward (1994) have developed a modelling framework for longitudinal data with informative dropouts, in which random or completely random dropout mechanisms are also included as explicit models. The essential feature of the procedure is a logistic regression model for the probability of dropping out, in which the explanatory variables can include previous values of the response variable, and, in addition, the *unobserved* value at dropout as a *latent* variable (i.e., an unobserved variable). In other words, the dropout probability is allowed to depend on both the *observed* measurement history and the unobserved value at dropout. This allows both a formal assessment of the type of dropout mechanism in the data, and the estimation of effects of interest, for example, treatment effects under different assumptions about the dropout mechanism. A full technical account of the model is given in Diggle and Kenward (1994) and a detailed example that uses the approach is described in Carpenter et al. (2002).

One of the problems for an investigator struggling to identify the dropout mechanism in a data set is that there are no routine methods to help, although a number of largely ad hoc graphical procedures can be used as described in Diggle (1998), Everitt (2002b) and Carpenter et al. (2002).

## 10.7 Summary

Linear mixed effects models are extremely useful for modelling longitudinal data. The models allow the correlations between the repeated measurements to be accounted for so that correct inferences can be drawn about the effects of covariates of interest on the repeated response values. In this chapter we have concentrated on responses that are continuous and conditional on the explanatory variables and random effects have a normal distribution. But random effects models can also be applied to non-normal responses, for example binary variables – see, for example, Everitt (2002b).

The lack of independence of repeated measures data is what makes the modelling of such data a challenge. But even when only a single measurement of a response is involved, correlation can, in some circumstances, occur between the response values of different individuals and cause similar problems. As an example consider a randomised clinical trial in which subjects are recruited at multiple study centres. The multicentre design can help to provide adequate sample sizes and enhance the generalisability of the results. However factors that vary by centre, including patient characteristics and medical practice patterns, may exert a sufficiently powerful effect to make inferences that ignore the 'clustering' seriously misleading. Consequently it may be necessary to incorporate random effects for centres into the analysis.

## Exercises

Ex. 10.1 Use the `lm` function to fit a model to the Beat the Blues data that assumes that the repeated measurements are independent. Compare the results to those from fitting the random intercept model `BtheB_lmer1`.

Ex. 10.2 Investigate whether there is any evidence of an interaction between treatment and time for the Beat the Blues data.

Ex. 10.3 Construct a plot of the mean profiles of both groups in the Beat the Blues data, showing also standard deviation bars at each time point.

Ex. 10.4 One very simple procedure for assessing the dropout mechanism suggested in Carpenter et al. (2002) involves plotting the observations for each treatment group, at each time point, differentiating between two categories of patients; those who do and those who do not attend their next scheduled visit. Any clear difference between the distributions of values for these two categories indicates that dropout is not completely at random. Produce such a plot for the Beat the Blues data.

Ex. 10.5 The `phosphate` data given in Table 10.2 show the plasma inorganic phosphate levels for 33 subjects, 20 of whom are controls and 13 of whom have been classified as obese (Davis, 2002). Produce separate plots of the profiles of the individuals in each group, and guided by these plots fit what you think might be sensible linear mixed effects models.

**Table 10.2**: `phosphate` data. Plasma inorganic phosphate levels for various time points after glucose challenge.

| group | t0 | t0.5 | t1 | t1.5 | t2 | t3 | t4 | t5 |
|---|---|---|---|---|---|---|---|---|
| control | 4.3 | 3.3 | 3.0 | 2.6 | 2.2 | 2.5 | 3.4 | 4.4 |
| control | 3.7 | 2.6 | 2.6 | 1.9 | 2.9 | 3.2 | 3.1 | 3.9 |
| control | 4.0 | 4.1 | 3.1 | 2.3 | 2.9 | 3.1 | 3.9 | 4.0 |
| control | 3.6 | 3.0 | 2.2 | 2.8 | 2.9 | 3.9 | 3.8 | 4.0 |
| control | 4.1 | 3.8 | 2.1 | 3.0 | 3.6 | 3.4 | 3.6 | 3.7 |
| control | 3.8 | 2.2 | 2.0 | 2.6 | 3.8 | 3.6 | 3.0 | 3.5 |
| control | 3.8 | 3.0 | 2.4 | 2.5 | 3.1 | 3.4 | 3.5 | 3.7 |
| control | 4.4 | 3.9 | 2.8 | 2.1 | 3.6 | 3.8 | 4.0 | 3.9 |
| control | 5.0 | 4.0 | 3.4 | 3.4 | 3.3 | 3.6 | 4.0 | 4.3 |
| control | 3.7 | 3.1 | 2.9 | 2.2 | 1.5 | 2.3 | 2.7 | 2.8 |
| control | 3.7 | 2.6 | 2.6 | 2.3 | 2.9 | 2.2 | 3.1 | 3.9 |
| control | 4.4 | 3.7 | 3.1 | 3.2 | 3.7 | 4.3 | 3.9 | 4.8 |
| control | 4.7 | 3.1 | 3.2 | 3.3 | 3.2 | 4.2 | 3.7 | 4.3 |
| control | 4.3 | 3.3 | 3.0 | 2.6 | 2.2 | 2.5 | 2.4 | 3.4 |
| control | 5.0 | 4.9 | 4.1 | 3.7 | 3.7 | 4.1 | 4.7 | 4.9 |
| control | 4.6 | 4.4 | 3.9 | 3.9 | 3.7 | 4.2 | 4.8 | 5.0 |
| control | 4.3 | 3.9 | 3.1 | 3.1 | 3.1 | 3.1 | 3.6 | 4.0 |
| control | 3.1 | 3.1 | 3.3 | 2.6 | 2.6 | 1.9 | 2.3 | 2.7 |
| control | 4.8 | 5.0 | 2.9 | 2.8 | 2.2 | 3.1 | 3.5 | 3.6 |
| control | 3.7 | 3.1 | 3.3 | 2.8 | 2.9 | 3.6 | 4.3 | 4.4 |
| obese | 5.4 | 4.7 | 3.9 | 4.1 | 2.8 | 3.7 | 3.5 | 3.7 |
| obese | 3.0 | 2.5 | 2.3 | 2.2 | 2.1 | 2.6 | 3.2 | 3.5 |
| obese | 4.9 | 5.0 | 4.1 | 3.7 | 3.7 | 4.1 | 4.7 | 4.9 |
| obese | 4.8 | 4.3 | 4.7 | 4.6 | 4.7 | 3.7 | 3.6 | 3.9 |
| obese | 4.4 | 4.2 | 4.2 | 3.4 | 3.5 | 3.4 | 3.8 | 4.0 |
| obese | 4.9 | 4.3 | 4.0 | 4.0 | 3.3 | 4.1 | 4.2 | 4.3 |
| obese | 5.1 | 4.1 | 4.6 | 4.1 | 3.4 | 4.2 | 4.4 | 4.9 |
| obese | 4.8 | 4.6 | 4.6 | 4.4 | 4.1 | 4.0 | 3.8 | 3.8 |
| obese | 4.2 | 3.5 | 3.8 | 3.6 | 3.3 | 3.1 | 3.5 | 3.9 |
| obese | 6.6 | 6.1 | 5.2 | 4.1 | 4.3 | 3.8 | 4.2 | 4.8 |
| obese | 3.6 | 3.4 | 3.1 | 2.8 | 2.1 | 2.4 | 2.5 | 3.5 |
| obese | 4.5 | 4.0 | 3.7 | 3.3 | 2.4 | 2.3 | 3.1 | 3.3 |
| obese | 4.6 | 4.4 | 3.8 | 3.8 | 3.8 | 3.6 | 3.8 | 3.8 |

*Source*: From Davis, C. S., *Statistical Methods for the Analysis of Repeated Measurements*, Springer, New York, 2002. With kind permission of Springer Science and Business Media.

CHAPTER 11

# Analysing Longitudinal Data II – Generalised Estimation Equations: Treating Respiratory Illness and Epileptic Seizures

## 11.1 Introduction

The data in Table 11.1 were collected in a clinical trial comparing two treatments for a respiratory illness (Davis, 1991).

**Table 11.1**: `respiratory` data. Randomised clinical trial data from patients suffering from respiratory illness. Only the data of the first four patients are shown here.

| centre | treatment | sex | age | status | month | subject |
|---:|---:|---:|---:|---:|---:|---:|
| 1 | placebo | female | 46 | poor | 0 | 1 |
| 1 | placebo | female | 46 | poor | 1 | 1 |
| 1 | placebo | female | 46 | poor | 2 | 1 |
| 1 | placebo | female | 46 | poor | 3 | 1 |
| 1 | placebo | female | 46 | poor | 4 | 1 |
| 1 | placebo | female | 28 | poor | 0 | 2 |
| 1 | placebo | female | 28 | poor | 1 | 2 |
| 1 | placebo | female | 28 | poor | 2 | 2 |
| 1 | placebo | female | 28 | poor | 3 | 2 |
| 1 | placebo | female | 28 | poor | 4 | 2 |
| 1 | treatment | female | 23 | good | 0 | 3 |
| 1 | treatment | female | 23 | good | 1 | 3 |
| 1 | treatment | female | 23 | good | 2 | 3 |
| 1 | treatment | female | 23 | good | 3 | 3 |
| 1 | treatment | female | 23 | good | 4 | 3 |
| 1 | placebo | female | 44 | good | 0 | 4 |
| 1 | placebo | female | 44 | good | 1 | 4 |
| 1 | placebo | female | 44 | good | 2 | 4 |
| 1 | placebo | female | 44 | good | 3 | 4 |
| 1 | placebo | female | 44 | poor | 4 | 4 |
| ⋮ | ⋮ | ⋮ | ⋮ | ⋮ | ⋮ | ⋮ |

In each of two centres, eligible patients were randomly assigned to active treatment or placebo. During the treatment, the respiratory status (categorised `poor` or `good`) was determined at each of four, monthly visits. The trial recruited 111 participants (54 in the active group, 57 in the placebo group) and there were no missing data for either the responses or the covariates. The question of interest is to assess whether the treatment is effective and to estimate its effect.

**Table 11.2**: `epilepsy` data. Randomised clinical trial data from patients suffering from epilepsy. Only the data of the first four patients are shown here.

| treatment | base | age | seizure.rate | period | subject |
|---|---|---|---|---|---|
| placebo | 11 | 31 | 5 | 1 | 1 |
| placebo | 11 | 31 | 3 | 2 | 1 |
| placebo | 11 | 31 | 3 | 3 | 1 |
| placebo | 11 | 31 | 3 | 4 | 1 |
| placebo | 11 | 30 | 3 | 1 | 2 |
| placebo | 11 | 30 | 5 | 2 | 2 |
| placebo | 11 | 30 | 3 | 3 | 2 |
| placebo | 11 | 30 | 3 | 4 | 2 |
| placebo | 6 | 25 | 2 | 1 | 3 |
| placebo | 6 | 25 | 4 | 2 | 3 |
| placebo | 6 | 25 | 0 | 3 | 3 |
| placebo | 6 | 25 | 5 | 4 | 3 |
| placebo | 8 | 36 | 4 | 1 | 4 |
| placebo | 8 | 36 | 4 | 2 | 4 |
| placebo | 8 | 36 | 1 | 3 | 4 |
| placebo | 8 | 36 | 4 | 4 | 4 |
| ⋮ | ⋮ | ⋮ | ⋮ | ⋮ | ⋮ |

In a clinical trial reported by Thall and Vail (1990), 59 patients with epilepsy were randomised to groups receiving either the antiepileptic drug Progabide or a placebo in addition to standard chemotherapy. The numbers of seizures suffered in each of four, two-week periods were recorded for each patient along with a baseline seizure count for the 8 weeks prior to being randomised to treatment and age. The main question of interest is whether taking Progabide reduced the number of epileptic seizures compared with placebo. A subset of the data is given in Table 11.2.

Note that the two data sets are shown in their 'long form' i.e., one measurement per row in the corresponding *data.frames*.

## 11.2 Generalised Estimating Equations

The data sets `respiratory` and `epilepsy` arise from longitudinal clinical trials, the same type of study that was the subject of consideration in Chapter 10. But in each case the repeatedly measured response variable is clearly not normally distributed making the models considered in the previous chapter unsuitable. Generalised linear mixed models could be applied to such non-normal response variables, for example using the function `glmmPQL` from package *MASS* (Venables and Ripley, 2002), but we will focus on another class of models suitable for correlated measurements in this chapter.

In Table 11.1 we have a binary response observed on four occasions, and in Table 11.2 a count response also observed on four occasions. If we choose to ignore the repeated measurements aspects of the two data sets we could use the methods of Chapter 6 applied to the data arranged in the 'long' form introduced in Chapter 10. For the `respiratory` data in Table 11.1 we could then apply logistic regression and for `epilepsy` in Table 11.2, Poisson regression. It can be shown that this approach will give *consistent* estimates of the regression coefficients, i.e., with large samples these point estimates should be close to the true population values. But the assumption of the independence of the repeated measurements will lead to estimated standard errors that are too small for the between-subjects covariates (at least when the correlation between the repeated measurements are positive) as a result of assuming that there are more independent data points than are justified.

We might begin by asking is there something relatively simple that can be done to 'fix-up' these standard errors so that we can still apply the R `glm` function to get reasonably satisfactory results on longitudinal data with a non-normal response? Two approaches which can often help to get more suitable estimates of the required standard errors are *bootstrapping* and use of the *robust/sandwich, Huber/White variance estimator*.

The idea underlying the bootstrap (see Chapters 7 and 8), a technique described in detail in Efron and Tibshirani (1993), is to resample from the observed data with replacement to achieve a sample of the same size each time, and to use the variation in the estimated parameters across the set of bootstrap samples in order to get a value for the sampling variability of the estimate (see Chapter 7 also). With correlated data, the bootstrap sample needs to be drawn with replacement from the set of independent subjects, so that intra-subject correlation is preserved in the bootstrap samples. We shall not consider this approach any further here.

The sandwich or robust estimate of variance (see Everitt and Pickles, 2000, for complete details including an explicit definition), involves, unlike the bootstrap which is computationally intensive, a closed-form calculation, based on an asymptotic (large-sample) approximation; it is known to provide good results in many situations. We shall illustrate its use in later examples.

But perhaps more satisfactory than these methods to simply 'fix-up' the standard errors given by the independence model, would be an approach that fully utilises information on the data's structure, including dependencies over time. A suitable procedure was first suggested by Liang and Zeger (1986) and is known as *generalised estimating equations* (GEE). In essence GEE is a multivariate extension of the generalised linear model and quasi-likelihood methods outlined in Chapter 6. The use of the latter leads to consistent inferences about mean responses without requiring specific assumptions to be made about second and higher order moments, thus avoiding intractable likelihood functions with possibly many nuisance parameters. Full details of the method are given in Liang and Zeger (1986) and Zeger and Liang (1986) but the primary idea behind the GEE approach is that since the parameters specifying the structure of the correlation matrix are rarely of great practical interest, simple structures are used for the within-subject correlations giving rise to the so-called *working correlation matrix*. Liang and Zeger (1986) show that the estimates of the parameters of most interest, i.e., those that determine the average responses over time, are still valid even when the correlation structure is incorrectly specified, although their standard errors might remain poorly estimated if the working correlation matrix is far from the truth. But as with the independence situation described previously, this potential difficulty can often be handled satisfactorily by again using the *sandwich estimator* to find more reasonable standard errors. Possibilities for the working correlation matrix that are most frequently used in practice are:

**An identity matrix** corresponds to what is usually termed the *independence working model*. Repeated responses are naively assumed to be independent, and standard generalised linear modelling software can be used for estimation etc., with each individual contributing as many records as repeated measures. Although clearly not realistic for most situations, used in association with a sandwich estimator of standard errors, it can lead to sensible inferences and has the distinct advantage of being simple to implement.

**An exchangeable correlation matrix** with a single parameter which gives the correlation of each pair of repeated measures. This assumption leads to the so-called *compound symmetry* structure for the covariance matrix of these measures, as described in Chapter 10.

**An autoregressive correlation matrix** also with a single parameter but in which $\text{corr}(y_j, y_k) = \vartheta^{|k-j|}, j \neq k$. With $\vartheta$ less than one this gives a pattern in which repeated measures further apart in time are less correlated, than those that are closer to one another.

**An unstructured correlation matrix** with $q(q-1)/2$ parameters in which $\text{corr}(y_j, y_k) = \vartheta_{jk}$ and where $q$ is the number of repeated measures. In general, using this form is not attractive because of the excess of number of parameters involved.

## 11.3 Analysis Using R

### *11.3.1 Beat the Blues Revisited*

Although we have introduced GEE as a method for analysing longitudinal data where the response variable is non-normal, it can also be applied to data where the response can be assumed to follow a conditional normal distribution (conditioning being on the explanatory variables). Consequently we first apply the method to the data used in the previous chapter so we can compare the results we get with those obtained from using the mixed-effects models used there.

To use the gee function, package *gee* (Carey et al., 2002) has to be installed and attached:

```
R> library("gee")
```

The gee function is used in a similar way to the lme function met in Chapter 10, with the addition of the features of the glm function that specify the appropriate error distribution for the response and the implied link function, and an argument to specify the structure of the working correlation matrix. Here we will fit an independence structure and then an exchangeable structure. The R code for fitting generalised estimation equations to the BtheB_long data (as constructed in Chapter 10) with idenity working correlation matrix is as follows (note that the gee function assumes the rows of the *data.frame* BtheB_long to be ordered with respect to subjects)

```
R> osub <- order(as.integer(BtheB_long$subject))
R> BtheB_long <- BtheB_long[osub, ]
R> btb_gee <- gee(bdi ~ bdi.pre + treatment + length +
+      drug, data = BtheB_long, id = subject, family = gaussian,
+      corstr = "independence")
```

and with exchangeable correlation matrix

```
R> btb_gee1 <- gee(bdi ~ bdi.pre + treatment + length +
+      drug, data = BtheB_long, id = subject, family = gaussian,
+      corstr = "exchangeable")
```

The summary method can be used to inspect the fitted models; the results are shown in Figures 11.1 and 11.2

Note how the naïve and the sandwich or robust estimates of the standard errors are considerably different for the independence structure (Figure 11.1), but quite similar for the exchangeable structure (Figure 11.2). This simply reflects that using an exchangeable working correlation matrix is more realistic for these data and that the standard errors resulting from this assumption are already quite reasonable without applying the 'sandwich' procedure to them. And if we compare the results under this assumed structure with those for the random intercept model given in Chapter 10 (Figure 10.2) we see that they are almost identical, since the random intercept model also implies an exchangeable structure for the correlations of the repeated measurements.

```
R> summary(btb_gee)

 GEE:  GENERALIZED LINEAR MODELS FOR DEPENDENT DATA
 gee S-function, version 4.13 modified 98/01/27 (1998)

Model:
 Link:                      Identity
 Variance to Mean Relation: Gaussian
 Correlation Structure:     Independent

Call:
gee(formula = bdi ~ bdi.pre + treatment + length + drug,
    id = subject, data = BtheB_long, family = gaussian,
    corstr = "independence")

Summary of Residuals:
        Min          1Q      Median          3Q         Max
-21.6497810  -5.8485100   0.1131663   5.5838383  28.1871039


Coefficients:
                 Estimate Naive S.E.   Naive z Robust S.E.
(Intercept)     3.5686314  1.4833349  2.405816  2.26947617
bdi.pre         0.5818494  0.0563904 10.318235  0.09156455
treatmentBtheB -3.2372285  1.1295569 -2.865928  1.77459534
length>6m       1.4577182  1.1380277  1.280916  1.48255866
drugYes        -3.7412982  1.1766321 -3.179667  1.78271179
                 Robust z
(Intercept)     1.5724472
bdi.pre         6.3545274
treatmentBtheB -1.8242066
length>6m       0.9832449
drugYes        -2.0986557

Estimated Scale Parameter:  79.25813
Number of Iterations:  1

Working Correlation
     [,1] [,2] [,3] [,4]
[1,]    1    0    0    0
[2,]    0    1    0    0
[3,]    0    0    1    0
[4,]    0    0    0    1
```

**Figure 11.1** R output of the `summary` method for the `btb_gee` model.

```
R> summary(btb_gee1)

 GEE:  GENERALIZED LINEAR MODELS FOR DEPENDENT DATA
 gee S-function, version 4.13 modified 98/01/27 (1998)

Model:
 Link:                      Identity
 Variance to Mean Relation: Gaussian
 Correlation Structure:     Exchangeable

Call:
gee(formula = bdi ~ bdi.pre + treatment + length + drug,
    id = subject, data = BtheB_long, family = gaussian,
    corstr = "exchangeable")

Summary of Residuals:
       Min         1Q     Median         3Q        Max
-23.955980  -6.643864  -1.109741   4.257688  25.452310


Coefficients:
                 Estimate Naive S.E.     Naive z Robust S.E.
(Intercept)     3.0231602 2.30390185  1.31219140  2.23204410
bdi.pre         0.6479276 0.08228567  7.87412417  0.08351405
treatmentBtheB -2.1692863 1.76642861 -1.22806339  1.73614385
length>6m      -0.1112910 1.73091679 -0.06429596  1.55092705
drugYes        -2.9995608 1.82569913 -1.64296559  1.73155411
                 Robust z
(Intercept)     1.3544357
bdi.pre         7.7583066
treatmentBtheB -1.2494854
length>6m      -0.0717577
drugYes        -1.7322940

Estimated Scale Parameter:  81.7349
Number of Iterations:  5

Working Correlation
          [,1]      [,2]      [,3]      [,4]
[1,] 1.0000000 0.6757951 0.6757951 0.6757951
[2,] 0.6757951 1.0000000 0.6757951 0.6757951
[3,] 0.6757951 0.6757951 1.0000000 0.6757951
[4,] 0.6757951 0.6757951 0.6757951 1.0000000
```

**Figure 11.2** R output of the `summary` method for the `btb_gee1` model.

The single estimated parameter for the working correlation matrix from the GEE procedure is 0.676, very similar to the estimated intra-class correlation coefficient from the random intercept model. i.e., $7.03^2/(5.07^2 + 7.03^2) = 0.66$ – see Figure 10.2.

### *11.3.2 Respiratory Illness*

We will now apply the GEE procedure to the `respiratory` data shown in Table 11.1. Given the binary nature of the response variable we will choose a binomial error distribution and by default a logistic link function. We shall also fix the scale parameter $\phi$ described in Chapter 6 at one. (The default in the `gee` function is to estimate this parameter.) Again we will apply the procedure twice, firstly with an independence structure and then with an exchangeable structure for the working correlation matrix. We will also fit a logistic regression model to the data using `glm` so we can compare results.

The baseline status, i.e., the status for `month == 0`, will enter the models as an explanatory variable and thus we have to rearrange the *data.frame* `respiratory` in order to create a new variable `baseline`:

```
R> data("respiratory", package = "HSAUR")
R> resp <- subset(respiratory, month > "0")
R> resp$baseline <- rep(subset(respiratory, month ==
+     "0")$status, rep(4, 111))
R> resp$nstat <- as.numeric(resp$status == "good")
```

The new variable `nstat` is simply a dummy coding for a poor respiratory status. Now we can use the data `resp` to fit a logistic regression model and GEE models with an independent and an exchangeable correlation structure as follows;

```
R> resp_glm <- glm(status ~ centre + treatment + sex +
+     baseline + age, data = resp, family = "binomial")
R> resp_gee1 <- gee(nstat ~ centre + treatment + sex +
+     baseline + age, data = resp, family = "binomial",
+     id = subject, corstr = "independence", scale.fix = TRUE,
+     scale.value = 1)
R> resp_gee2 <- gee(nstat ~ centre + treatment + sex +
+     baseline + age, data = resp, family = "binomial",
+     id = subject, corstr = "exchangeable", scale.fix = TRUE,
+     scale.value = 1)
```

Again, `summary` methods can be used for a inspection of the details of the fitted models, the results are given in Figures 11.3, 11.4 and 11.5. We see that the results from applying logistic regression to the data with the `glm` function gives identical results to those obtained from `gee` with an independence correlation structure (comparing the `glm` standard errors with the naïve standard errors from `gee`). The robust standard errors for the between subject covariates are considerably larger than those estimated assuming independence, implying that the independence assumption is not realistic for these

```
R> summary(resp_glm)

Call:
glm(formula = status ~ centre + treatment + sex + baseline +
    age, family = "binomial", data = resp)

Deviance Residuals:
    Min       1Q   Median       3Q      Max
-2.3146  -0.8551   0.4336   0.8953   1.9246

Coefficients:
                     Estimate Std. Error z value Pr(>|z|)
(Intercept)         -0.900171   0.337653  -2.666  0.00768 **
centre2              0.671601   0.239567   2.803  0.00506 **
treatmenttreatment   1.299216   0.236841   5.486 4.12e-08 ***
sexmale              0.119244   0.294671   0.405  0.68572
baselinegood         1.882029   0.241290   7.800 6.20e-15 ***
age                 -0.018166   0.008864  -2.049  0.04043 *
---
Signif. codes:  0 '***' 0.001 '**' 0.01 '*' 0.05 '.' 0.1 ' ' 1

(Dispersion parameter for binomial family taken to be 1)

    Null deviance: 608.93  on 443  degrees of freedom
Residual deviance: 483.22  on 438  degrees of freedom
AIC: 495.22

Number of Fisher Scoring iterations: 4
```

**Figure 11.3** R output of the summary method for the resp_glm model.

data. Applying the GEE procedure with an exchangeable correlation structure results in naïve and robust standard errors that are identical, and similar to the robust estimates from the independence structure. It is clear that the exchangeable structure more adequately reflects the correlational structure of the observed repeated measurements than does independence.

```
R> summary(resp_gee1)

 GEE:  GENERALIZED LINEAR MODELS FOR DEPENDENT DATA
 gee S-function, version 4.13 modified 98/01/27 (1998)

Model:
 Link:                      Logit
 Variance to Mean Relation: Binomial
 Correlation Structure:     Independent

Call:
gee(formula = nstat ~ centre + treatment + sex + baseline +
    age, id = subject, data = resp, family = "binomial",
    corstr = "independence", scale.fix = TRUE,
    scale.value = 1)

Summary of Residuals:
        Min          1Q      Median          3Q         Max
-0.93134415 -0.30623174  0.08973552  0.33018952  0.84307712


Coefficients:
                      Estimate  Naive S.E.   Naive z
(Intercept)        -0.90017133 0.337653052 -2.665965
centre2             0.67160098 0.239566599  2.803400
treatmenttreatment  1.29921589 0.236841017  5.485603
sexmale             0.11924365 0.294671045  0.404667
baselinegood        1.88202860 0.241290221  7.799854
age                -0.01816588 0.008864403 -2.049306
                   Robust S.E.   Robust z
(Intercept)         0.46032700 -1.9555041
centre2             0.35681913  1.8821889
treatmenttreatment  0.35077797  3.7038127
sexmale             0.44320235  0.2690501
baselinegood        0.35005152  5.3764332
age                 0.01300426 -1.3969169

Estimated Scale Parameter:  1
Number of Iterations:  1

Working Correlation
     [,1] [,2] [,3] [,4]
[1,]    1    0    0    0
[2,]    0    1    0    0
[3,]    0    0    1    0
[4,]    0    0    0    1
```

**Figure 11.4** R output of the summary method for the resp_gee1 model.

```
R> summary(resp_gee2)

 GEE:  GENERALIZED LINEAR MODELS FOR DEPENDENT DATA
 gee S-function, version 4.13 modified 98/01/27 (1998)

Model:
 Link:                      Logit
 Variance to Mean Relation: Binomial
 Correlation Structure:     Exchangeable

Call:
gee(formula = nstat ~ centre + treatment + sex + baseline +
    age, id = subject, data = resp, family = "binomial",
    corstr = "exchangeable", scale.fix = TRUE,
    scale.value = 1)

Summary of Residuals:
        Min          1Q      Median          3Q         Max
-0.93134415 -0.30623174  0.08973552  0.33018952  0.84307712


Coefficients:
                       Estimate Naive S.E.    Naive z
(Intercept)         -0.90017133 0.47846344 -1.8813796
centre2              0.67160098 0.33947230  1.9783676
treatmenttreatment   1.29921589 0.33561008  3.8712064
sexmale              0.11924365 0.41755678  0.2855747
baselinegood         1.88202860 0.34191472  5.5043802
age                 -0.01816588 0.01256110 -1.4462014
                    Robust S.E.   Robust z
(Intercept)          0.46032700 -1.9555041
centre2              0.35681913  1.8821889
treatmenttreatment   0.35077797  3.7038127
sexmale              0.44320235  0.2690501
baselinegood         0.35005152  5.3764332
age                  0.01300426 -1.3969169

Estimated Scale Parameter:  1
Number of Iterations:  1

Working Correlation
          [,1]      [,2]      [,3]      [,4]
[1,] 1.0000000 0.3359883 0.3359883 0.3359883
[2,] 0.3359883 1.0000000 0.3359883 0.3359883
[3,] 0.3359883 0.3359883 1.0000000 0.3359883
[4,] 0.3359883 0.3359883 0.3359883 1.0000000
```

**Figure 11.5** R output of the `summary` method for the `resp_gee2` model.

The estimated treatment effect taken from the exchangeable structure GEE model is 1.299 which, using the robust standard errors, has an associated 95% confidence interval

```
R> se <- summary(resp_gee2)$coefficients["treatmenttreatment",
+     "Robust S.E."]
R> coef(resp_gee2)["treatmenttreatment"] + c(-1, 1) *
+     se * qnorm(0.975)

[1] 0.6117037 1.9867281
```

These values reflect effects on the log-odds scale. Interpretation becomes simpler if we exponentiate the values to get the effects in terms of odds. This gives a treatment effect of 3.666 and a 95% confidence interval of

```
R> exp(coef(resp_gee2)["treatmenttreatment"] + c(-1,
+     1) * se * qnorm(0.975))

[1] 1.843570 7.291637
```

The odds of achieving a 'good' respiratory status with the active treatment is between about twice and seven times the corresponding odds for the placebo.

### 11.3.3 Epilepsy

Moving on to the count data in `epilepsy` from Table 11.2, we begin by calculating the means and variances of the number of seizures for all treatment / period interactions

```
R> data("epilepsy", package = "HSAUR")
R> itp <- interaction(epilepsy$treatment, epilepsy$period)
R> tapply(epilepsy$seizure.rate, itp, mean)

  placebo.1 Progabide.1   placebo.2 Progabide.2   placebo.3
   9.357143    8.580645    8.285714    8.419355    8.785714
Progabide.3   placebo.4 Progabide.4
   8.129032    7.964286    6.709677

R> tapply(epilepsy$seizure.rate, itp, var)

  placebo.1 Progabide.1   placebo.2 Progabide.2   placebo.3
  102.75661   332.71828    66.65608   140.65161   215.28571
Progabide.3   placebo.4 Progabide.4
  193.04946    58.18386   126.87957
```

Some of the variances are considerably larger than the corresponding means, which for a Poisson variable may suggest that overdispersion may be a problem, see Chapter 6.

We will now construct some boxplots first for the numbers of seizures observed in each two-week period post randomisation. The resulting diagram is shown in Figure 11.6. Some quite extreme 'outliers' are indicated, particularly the observation in period one in the Progabide group. But given these are count data which we will model using a Poisson error distribution and a

```
R> layout(matrix(1:2, nrow = 1))
R> ylim <- range(epilepsy$seizure.rate)
R> placebo <- subset(epilepsy, treatment == "placebo")
R> progabide <- subset(epilepsy, treatment == "Progabide")
R> boxplot(seizure.rate ~ period, data = placebo,
+     ylab = "Number of seizures",
+     xlab = "Period", ylim = ylim, main = "Placebo")
R> boxplot(seizure.rate ~ period, data = progabide,
+     main = "Progabide", ylab = "Number of seizures",
+     xlab = "Period", ylim = ylim)
```

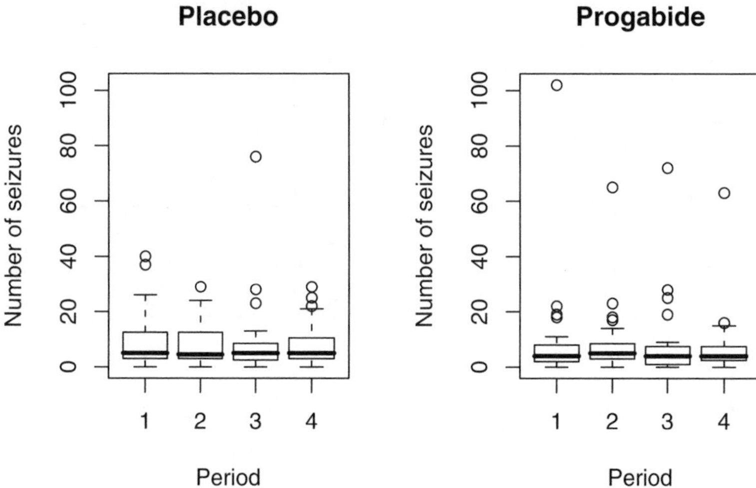


**Figure 11.6** Boxplots of numbers of seizures in each two-week period post randomisation for placebo and active treatments.

log link function, it may be more appropriate to look at the boxplots *after* taking a log transformation. (Since some observed counts are zero we will add 1 to all observations before taking logs.) To get the plots we can use the R code displayed with Figure 11.7. In Figure 11.7 the outlier problem seems less troublesome and we shall not attempt to remove any of the observations for subsequent analysis.

Before proceeding with the formal analysis of these data we have to deal with a small problem produced by the fact that the baseline counts were observed over an eight-week period whereas all subsequent counts are over two-week periods. For the baseline count we shall simply divide by eight to get an average weekly rate, but we cannot do the same for the post-randomisation counts if we are going to assume a Poisson distribution (since we will no longer have

```
R> layout(matrix(1:2, nrow = 1))
R> ylim <- range(log(epilepsy$seizure.rate + 1))
R> boxplot(log(seizure.rate + 1) ~ period, data = placebo,
+      main = "Placebo", ylab = "Log number of seizures",
+      xlab = "Period", ylim = ylim)
R> boxplot(log(seizure.rate + 1) ~ period, data = progabide,
+      main = "Progabide", ylab = "Log number of seizures",
+      xlab = "Period", ylim = ylim)
```

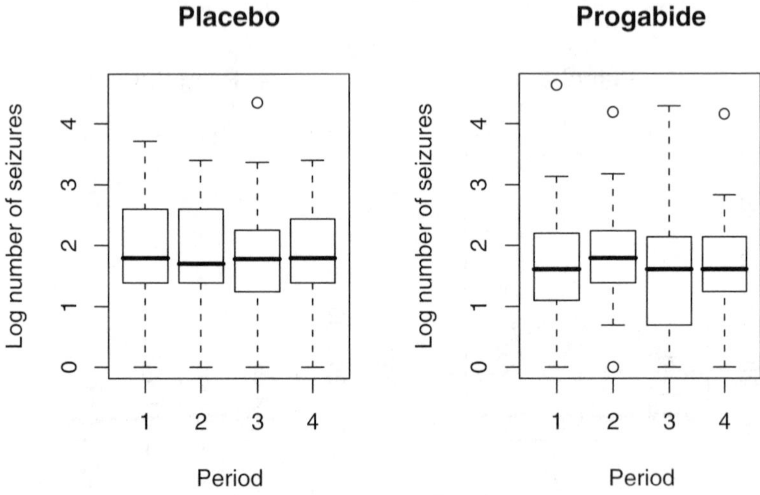


**Figure 11.7** Boxplots of log of numbers of seizures in each two-week period post randomisation for placebo and active treatments.

integer values for the response). But we can model the mean count for each two-week period by introducing the log of the observation period as an *offset* (a covariate with regression coefficient set to one). The model then becomes log(expected count in observation period) = linear function of explanatory variables+log(observation period), leading to the model for the rate in counts per week (assuming the observation periods are measured in weeks) as expected count in observation period/observation period = exp(linear function of explanatory variables). In our example the observation period is two weeks, so we simply need to set log(2) for each observation as the offset.

We can now fit a Poisson regression model to the data assuming independence using the `glm` function. We also use the GEE approach to fit an independence structure, followed by an exchangeable structure using the following R code:

```
R> per <- rep(log(2), nrow(epilepsy))
R> epilepsy$period <- as.numeric(epilepsy$period)
R> epilepsy_glm <- glm(seizure.rate ~ base + age +
+     treatment + offset(per), data = epilepsy,
+     family = "poisson")
R> epilepsy_gee1 <- gee(seizure.rate ~ base + age +
+     treatment + offset(per), data = epilepsy,
+     family = "poisson",
+     id = subject, corstr = "independence", scale.fix = TRUE,
+     scale.value = 1)
R> epilepsy_gee2 <- gee(seizure.rate ~ base + age +
+     treatment + offset(per), data = epilepsy,
+     family = "poisson",
+     id = subject, corstr = "exchangeable", scale.fix = TRUE,
+     scale.value = 1)
R> epilepsy_gee3 <- gee(seizure.rate ~ base + age +
+     treatment + offset(per), data = epilepsy,
+     family = "poisson",
+     id = subject, corstr = "exchangeable", scale.fix = FALSE,
+     scale.value = 1)
```

As usual we inspect the fitted models using the `summary` method, the results are given in Figures 11.8, 11.9, 11.10, and 11.11.

For this example, the estimates of standard errors under independence are about half of the corresponding robust estimates, and the situation improves only a little when an exchangeable structure is fitted. Using the naïve standard errors leads, in particular, to a highly significant treatment effect which disappears when the robust estimates are used. The problem with the GEE approach here, using either the independence or exchangeable correlation structure lies in constraining the scale parameter to be one. For these data there is overdispersion which has to be accommodated by allowing this parameter to be freely estimated. When this is done, it gives the last set of results shown above. The estimate of $\phi$ is 5.09 and the naïve and robust estimates of the standard errors are now very similar. It is clear that there is no evidence of a treatment effect.

```
R> summary(epilepsy_glm)

Call:
glm(formula = seizure.rate ~ base + age + treatment +
    offset(per), family = "poisson", data = epilepsy)

Deviance Residuals:
    Min       1Q   Median       3Q      Max
-4.4360  -1.4034  -0.5029   0.4842  12.3223

Coefficients:
                      Estimate Std. Error z value Pr(>|z|)
(Intercept)         -0.1306156  0.1356191  -0.963  0.33549
base                 0.0226517  0.0005093  44.476  < 2e-16 ***
age                  0.0227401  0.0040240   5.651 1.59e-08 ***
treatmentProgabide -0.1527009  0.0478051  -3.194  0.00140 **
---
Signif. codes:  0 '***' 0.001 '**' 0.01 '*' 0.05 '.' 0.1 ' ' 1

(Dispersion parameter for poisson family taken to be 1)

    Null deviance: 2521.75  on 235  degrees of freedom
Residual deviance:  958.46  on 232  degrees of freedom
AIC: 1732.5

Number of Fisher Scoring iterations: 5
```

**Figure 11.8** R output of the summary method for the epilepsy_glm model.

```
R> summary(epilepsy_gee1)

 GEE:  GENERALIZED LINEAR MODELS FOR DEPENDENT DATA
 gee S-function, version 4.13 modified 98/01/27 (1998)

Model:
 Link:                      Logarithm
 Variance to Mean Relation: Poisson
 Correlation Structure:     Independent

Call:
gee(formula = seizure.rate ~ base + age + treatment +
    offset(per), id = subject, data = epilepsy,
    family = "poisson", corstr = "independence",
    scale.fix = TRUE, scale.value = 1)

Summary of Residuals:
       Min         1Q     Median         3Q        Max
-4.9195387  0.1808059  1.7073405  4.8850644 69.9658560


Coefficients:
                      Estimate   Naive S.E.    Naive z
(Intercept)        -0.13061561 0.1356191185 -0.9631062
base                0.02265174 0.0005093011 44.4761250
age                 0.02274013 0.0040239970  5.6511312
treatmentProgabide -0.15270095 0.0478051054 -3.1942393
                   Robust S.E.   Robust z
(Intercept)        0.365148155 -0.3577058
base               0.001235664 18.3316325
age                0.011580405  1.9636736
treatmentProgabide 0.171108915 -0.8924196

Estimated Scale Parameter:  1
Number of Iterations:  1

Working Correlation
     [,1] [,2] [,3] [,4]
[1,]    1    0    0    0
[2,]    0    1    0    0
[3,]    0    0    1    0
[4,]    0    0    0    1
```

**Figure 11.9** R output of the summary method for the epilepsy_gee1 model.

```
R> summary(epilepsy_gee2)

 GEE:  GENERALIZED LINEAR MODELS FOR DEPENDENT DATA
 gee S-function, version 4.13 modified 98/01/27 (1998)

Model:
 Link:                      Logarithm
 Variance to Mean Relation: Poisson
 Correlation Structure:     Exchangeable

Call:
gee(formula = seizure.rate ~ base + age + treatment +
    offset(per), id = subject, data = epilepsy,
    family = "poisson", corstr = "exchangeable",
    scale.fix = TRUE, scale.value = 1)

Summary of Residuals:
       Min         1Q     Median         3Q        Max
-4.9195387  0.1808059  1.7073405  4.8850644 69.9658560


Coefficients:
                      Estimate   Naive S.E.   Naive z
(Intercept)        -0.13061561 0.2004416507 -0.651639
base                0.02265174 0.0007527342 30.092612
age                 0.02274013 0.0059473665  3.823564
treatmentProgabide -0.15270095 0.0706547450 -2.161227
                   Robust S.E.   Robust z
(Intercept)        0.365148155 -0.3577058
base               0.001235664 18.3316325
age                0.011580405  1.9636736
treatmentProgabide 0.171108915 -0.8924196

Estimated Scale Parameter:  1
Number of Iterations:  1

Working Correlation
          [,1]      [,2]      [,3]      [,4]
[1,] 1.0000000 0.3948033 0.3948033 0.3948033
[2,] 0.3948033 1.0000000 0.3948033 0.3948033
[3,] 0.3948033 0.3948033 1.0000000 0.3948033
[4,] 0.3948033 0.3948033 0.3948033 1.0000000
```

**Figure 11.10** R output of the `summary` method for the `epilepsy_gee2` model.

```
R> summary(epilepsy_gee3)

 GEE:  GENERALIZED LINEAR MODELS FOR DEPENDENT DATA
 gee S-function, version 4.13 modified 98/01/27 (1998)

Model:
 Link:                      Logarithm
 Variance to Mean Relation: Poisson
 Correlation Structure:     Exchangeable

Call:
gee(formula = seizure.rate ~ base + age + treatment +
    offset(per), id = subject, data = epilepsy,
    family = "poisson", corstr = "exchangeable",
    scale.fix = FALSE, scale.value = 1)

Summary of Residuals:
      Min         1Q     Median          3Q         Max
-4.9195387  0.1808059  1.7073405  4.8850644 69.9658560


Coefficients:
                       Estimate  Naive S.E.    Naive z
(Intercept)         -0.13061561 0.452199543 -0.2888451
base                 0.02265174 0.001698180 13.3388301
age                  0.02274013 0.013417353  1.6948302
treatmentProgabide  -0.15270095 0.159398225 -0.9579840
                    Robust S.E.   Robust z
(Intercept)         0.365148155 -0.3577058
base                0.001235664 18.3316325
age                 0.011580405  1.9636736
treatmentProgabide  0.171108915 -0.8924196

Estimated Scale Parameter:  5.089608
Number of Iterations:  1

Working Correlation
          [,1]      [,2]      [,3]      [,4]
[1,] 1.0000000 0.3948033 0.3948033 0.3948033
[2,] 0.3948033 1.0000000 0.3948033 0.3948033
[3,] 0.3948033 0.3948033 1.0000000 0.3948033
[4,] 0.3948033 0.3948033 0.3948033 1.0000000
```

**Figure 11.11** R output of the summary method for the epilepsy_gee3 model.

## 11.4 Summary

The generalised estimation equation approach essentially extends generalised linear models to longitudinal data, and allows for the analysis of such data when the response variable cannot be assumed to be normal distributed.

### Exercises

Ex. 11.1 For the `epilepsy` data investigate what Poisson models are most suitable when subject 49 is excluded from the analysis.

Ex. 11.2 Investigate the use of other correlational structures than the independence and exchangeable structures used in the text, for both the `respiratory` and the `epilepsy` data.

Ex. 11.3 The data shown in Table 11.3 were collected in a follow-up study of women patients with schizophrenia (Davis, 2002). The binary response recorded at 0, 2, 6, 8 and 10 months after hospitalisation was thought disorder (absent or present). The single covariate is the factor indicating whether a patient had suffered early or late onset of her condition (age of onset less than 20 years or age of onset 20 years or above). The question of interest is whether the course of the illness differs between patients with early and late onset? Investigate this question using the GEE approach.

**Table 11.3**: `schizophrenia2` data. Clinical trial data from patients suffering from schizophrenia. Only the data of the first four patients are shown here.

| subject | onset | disorder | month |
|---:|---|---:|---:|
| 1 | $<$ 20 yrs | present | 0 |
| 1 | $<$ 20 yrs | present | 2 |
| 1 | $<$ 20 yrs | absent | 6 |
| 1 | $<$ 20 yrs | absent | 8 |
| 1 | $<$ 20 yrs | absent | 10 |
| 2 | $>$ 20 yrs | absent | 0 |
| 2 | $>$ 20 yrs | absent | 2 |
| 2 | $>$ 20 yrs | absent | 6 |
| 2 | $>$ 20 yrs | absent | 8 |
| 2 | $>$ 20 yrs | absent | 10 |
| 3 | $<$ 20 yrs | present | 0 |
| 3 | $<$ 20 yrs | present | 2 |
| 3 | $<$ 20 yrs | absent | 6 |
| 3 | $<$ 20 yrs | absent | 8 |
| 3 | $<$ 20 yrs | absent | 10 |
| 4 | $<$ 20 yrs | absent | 0 |
| 4 | $<$ 20 yrs | absent | 2 |
| 4 | $<$ 20 yrs | absent | 6 |
| 4 | $<$ 20 yrs | absent | 8 |
| 4 | $<$ 20 yrs | absent | 10 |
| ⋮ | ⋮ | ⋮ | ⋮ |

*Source*: From Davis, C. S., *Statistical Methods for the Analysis of Repeated Measurements*, Springer, New York, 2002. With kind permission of Springer Science and Business Media.

CHAPTER 12

# Meta-Analysis: Nicotine Gum and Smoking Cessation and the Efficacy of BCG Vaccine in the Treatment of Tuberculosis

## 12.1 Introduction

Cigarette smoking is the leading cause of preventable death in the United States and kills more Americans than AIDS, alcohol, illegal drug use, car accidents, fires, murders and suicides combined. It has been estimated that 430,000 Americans die from smoking every year. Fighting tobacco use is, consequently, one of the major public health goals of our time and there are now many programs available designed to help smokers quit. One of the major aids used in these programs is nicotine chewing gum, which acts as a substitute oral activity and provides a source of nicotine that reduces the withdrawal symptoms experienced when smoking is stopped. But separate randomised clinical trials of nicotine gum have been largely inconclusive, leading Silagy (2003) to consider combining the results from 26 such studies found from an extensive literature search. The results of these trials in terms of numbers of people in the treatment arm and the control arm who stopped smoking for at least 6 months after treatment are given in Table 12.1.

Bacille Calmette Guerin (BCG) is the most widely used vaccination in the world. Developed in the 1930s and made of a live, weakened strain of Mycobacterium bovis, the BCG is the only vaccination available against tuberculosis (TBC) today. Colditz et al. (1994) report data from 13 clinical trials of BCG vaccine each investigating its efficacy in the treatment of tuberculosis. The number of subjects suffering from TB with or without BCG vaccination are given in Table 12.2. In addition, the table contains the values of two other variables for each study, namely, the geographic latitude of the place where the study was undertaken and the year of publication. These two variables will be used to investigate and perhaps explain any heterogeneity among the studies.

**Table 12.1**: `smoking` data. Meta-analysis on nicotine gum showing the number of quitters who have been treated (`qt`), the total number of treated (`tt`) as well as the number of quitters in the control group (`qc`) with total number of smokers in the control group (`tc`).

| | `qt` | `tt` | `qc` | `tc` |
|---|---|---|---|---|
| Blondal89 | 37 | 92 | 24 | 90 |
| Campbell91 | 21 | 107 | 21 | 105 |
| Fagerstrom82 | 30 | 50 | 23 | 50 |
| Fee82 | 23 | 180 | 15 | 172 |
| Garcia89 | 21 | 68 | 5 | 38 |
| Garvey00 | 75 | 405 | 17 | 203 |
| Gross95 | 37 | 131 | 6 | 46 |
| Hall85 | 18 | 41 | 10 | 36 |
| Hall87 | 30 | 71 | 14 | 68 |
| Hall96 | 24 | 98 | 28 | 103 |
| Hjalmarson84 | 31 | 106 | 16 | 100 |
| Huber88 | 31 | 54 | 11 | 60 |
| Jarvis82 | 22 | 58 | 9 | 58 |
| Jensen91 | 90 | 211 | 28 | 82 |
| Killen84 | 16 | 44 | 6 | 20 |
| Killen90 | 129 | 600 | 112 | 617 |
| Malcolm80 | 6 | 73 | 3 | 121 |
| McGovern92 | 51 | 146 | 40 | 127 |
| Nakamura90 | 13 | 30 | 5 | 30 |
| Niaura94 | 5 | 84 | 4 | 89 |
| Pirie92 | 75 | 206 | 50 | 211 |
| Puska79 | 29 | 116 | 21 | 113 |
| Schneider85 | 9 | 30 | 6 | 30 |
| Tonnesen88 | 23 | 60 | 12 | 53 |
| Villa99 | 11 | 21 | 10 | 26 |
| Zelman92 | 23 | 58 | 18 | 58 |

**Table 12.2:** `BCG` data. Meta-analysis on BCG vaccine with the following data: the number of TBC cases after a vaccination with BCG (`BCGTB`), the total number of people who received BCG (`BCG`) as well as the number of TBC cases without vaccination (`NoVaccTB`) and the total number of people in the study without vaccination (`NoVacc`).

| Study | BCGTB | BCGVacc | NoVaccTB | NoVacc | Latitude | Year |
|---|---|---|---|---|---|---|
| 1 | 4 | 123 | 11 | 139 | 44 | 1948 |
| 2 | 6 | 306 | 29 | 303 | 55 | 1949 |
| 3 | 3 | 231 | 11 | 220 | 42 | 1960 |
| 4 | 62 | 13598 | 248 | 12867 | 52 | 1977 |
| 5 | 33 | 5069 | 47 | 5808 | 13 | 1973 |
| 6 | 180 | 1541 | 372 | 1451 | 44 | 1953 |
| 7 | 8 | 2545 | 10 | 629 | 19 | 1973 |
| 8 | 505 | 88391 | 499 | 88391 | 13 | 1980 |
| 9 | 29 | 7499 | 45 | 7277 | 27 | 1968 |
| 10 | 17 | 1716 | 65 | 1665 | 42 | 1961 |
| 11 | 186 | 50634 | 141 | 27338 | 18 | 1974 |
| 12 | 5 | 2498 | 3 | 2341 | 33 | 1969 |
| 13 | 27 | 16913 | 29 | 17854 | 33 | 1976 |

## 12.2 Systematic Reviews and Meta-Analysis

Many individual clinical trials are not large enough to answer the questions we want to answer as reliably as we would want to answer them. Often trials are too small for adequate conclusions to be drawn about potentially small advantages of particular therapies. Advocacy of large trials is a natural response to this situation, but it is not always possible to launch very large trials before therapies become widely accepted or rejected prematurely. One possible answer to this problem lies in the classical narrative review of a set of clinical trials with an accompanying informal synthesis of evidence from the different studies. It is now generally recognised however that such review articles can, unfortunately, be very misleading as a result of both the possible biased selection of evidence and the emphasis placed upon it by the reviewer to support his or her personal opinion.

An alternative approach that has become increasingly popular in the last decade or so is the *systematic review* which has, essentially, two components:

**Qualitative:** the description of the available trials, in terms of their relevance and methodological strengths and weaknesses.

**Quantitative:** a means of mathematically combining results from different studies, even when these studies have used different measures to assess the dependent variable.

The quantitative component of a systematic review is usually known as a *meta-analysis*, defined in the 'Cambridge Dictionary of Statistics in the Medical Sciences' (Everitt, 2002a), as follows:

> A collection of techniques whereby the results of two or more independent studies are statistically combined to yield an overall answer to a question of interest. The rationale behind this approach is to provide a test with more power than is provided by the separate studies themselves. The procedure has become increasingly popular in the last decade or so, but is not without its critics, particularly because of the difficulties of knowing which studies should be included and to which population final results actually apply.

It is now generally accepted that meta-analysis gives the systematic review an objectivity that is inevitably lacking in literature reviews and can also help the process to achieve greater precision and generalisability of findings than any single study. Chalmers and Lau (1993) make the point that both the classical review article and a meta-analysis can be biased, but that at least the writer of a meta-analytic paper is required by the rudimentary standards of the discipline to give the data on which any conclusions are based, and to defend the development of these conclusions by giving evidence that all available data are included, or to give the reasons for not including the data. Chalmers and Lau (1993) conclude;

> It seems obvious that a discipline that requires all available data be revealed and included in an analysis has an advantage over one that has traditionally not presented analyses of all the data in which conclusions are based.

The demand for systematic reviews of health care interventions has developed rapidly during the last decade, initiated by the widespread adoption of the principles of *evidence-based medicine* both amongst health care practitioners and policy makers. Such reviews are now increasingly used as a basis for both individual treatment decisions and the funding of health care and health care research worldwide. Systematic reviews have a number of aims:

- To review systematically the available evidence from a particular research area,
- To provide quantitative summaries of the results from each study,
- To combine the results across studies if appropriate; such combination of results leads to greater statistical power in estimating treatment effects,
- To assess the amount of variability between studies,
- To estimate the degree of benefit associated with a particular study treatment,
- To identify study characteristics associated with particularly effective treatments.

Perhaps the most important aspect of a meta-analysis is study selection. Selection is a matter of inclusion and exclusion and the judgements required are, at times, problematic. But we shall say nothing about this fundamental component of a meta-analysis here since it has been comprehensively dealt with by a number of authors, including Chalmers and Lau (1993) and Petitti (2000). Instead we shall concentrate on the statistics of meta-analysis.

## 12.3 Statistics of Meta-Analysis

Two models that are frequently used in the meta-analysis of medical studies are the *fixed effects* and *random effects* models. Whilst the former assumes that each observed individual study result is estimating a common, unknown overall pooled effect, the latter assumes that each individual observed result is estimating its own unknown underlying effect, which in turn are estimating a common population mean. Thus the random effects model specifically allows for the existence of *between study heterogeneity* as well as *within-study variability.* DeMets (1987) and Bailey (1987) discuss the strengths and weaknesses of the two competing models. Bailey suggests that when the research question involves extrapolation to the future – *will* the treatment have an effect, on the average – then the random effects model for the studies is the appropriate one. The research question implicitly assumes that there is a population of studies from which those analysed in the meta-analysis were sampled, and anticipate future studies being conducted or previously unknown studies being uncovered.

When the research question concerns whether treatment *has* produced an effect, on the average, *in the set of studies being analysed*, then the fixed effects model for the studies may be the appropriate one; here there is no interest in generalising the results to other studies.

Many statisticians believe that random effects models are more appropriate than fixed effects models for meta-analysis because between-study variation is an important source of uncertainty that should not be ignored.

### *12.3.1 Fixed Effects Model – Mantel-Haenszel*

This model uses as its estimate of the common pooled effect, $\bar{Y}$, a weighted average of the individual study effects, the weights being inversely proportional to the within-study variances. Specifically

$$\bar{Y} = \frac{\sum\limits_{i=1}^{k} W_i Y_i}{\sum\limits_{i=1}^{k} W_i} \tag{12.1}$$

where $k$ is the number of the studies in the meta-analysis, $Y_i$ is the effect size estimated in the $i$th study (this might be a log-odds ratio, relative risk or difference in means, for example), and $W_i = 1/V_i$ where $V_i$ is the within study estimate of variance for the $i$th study. The estimated variance of $\bar{Y}$ is given by

$$\mathsf{Var}(\bar{Y}) = 1/\left(\sum_{i=1}^{k} W_i\right). \qquad (12.2)$$

From (12.1) and (12.2) a confidence interval for the pooled effect can be constructed in the usual way.

### *12.3.2 Random Effects Model – DerSimonian-Laird*

The random effects model has the form;

$$\begin{aligned} Y_i &= \mu_i + \sigma_i \varepsilon_i; \quad \varepsilon_i \sim \mathcal{N}(0,1) \\ \mu_i &\sim N(\mu, \tau^2); \quad i = 1, \dots, k. \end{aligned} \qquad (12.3)$$

Unlike the fixed effects model, the individual studies are not assumed to be estimating a true single effect size; rather the true effects in each study, the $\mu_i$'s are assumed to have been sampled from a distribution of effects, assumed to be normal with mean $\mu$ and variance $\tau^2$. The estimate of $\mu$ is that given in (12.1) but in this case the weights are given by $W_i = 1/\left(V_i^2 + \hat{\tau}^2\right)$ where $\hat{\tau}^2$ is an estimate of the between study variance. DerSimonian and Laird (1986) derive a suitable estimator for $\hat{\tau}^2$, which is as follows;

$$\hat{\tau}^2 = \begin{cases} 0 & \text{if } Q < k-1 \\ (Q-k+1)/U & \text{if } Q > k-1 \end{cases}$$

where $Q = \sum_{i=1}^{k} W_i(Y_i - \bar{Y})^2$ and $U = (k-1)\left(\bar{W} - s_W^2/kW\right)$ with $\bar{W}$ and $s_W^2$ being the mean and variance of the weights, $W_i$.

A test for homogeneity of studies is provided by the statistic $Q$. The hypothesis of a common effect size is rejected if $Q$ exceeds the quantile of a $\chi^2$-distribution with $k-1$ degrees of freedom at the chosen significance level.

Allowing for this extra between-study variation has the effect of reducing the relative weighting given to the more precise studies. Hence the random effects model produces a more conservative confidence interval for the pooled effect size.

A Bayesian dimension can be added to the random effects model by allowing the parameters of the model to have prior distributions. Some examples are given in Sutton and Abrams (2001).

## 12.4 Analysis Using R

The methodology described above is implemented in package *rmeta* (Lumley, 2005) and we will utilise the functionality from this package to analyse the smoking and BCG studies introduced earlier.

The aim in collecting the results from the randomised trials of using nicotine gum to help smokers quit was to estimate the overall *odds ratio*, the odds of quitting smoking for those given the gum, divided by the odds of quitting for those not receiving the gum. The odds ratios and corresponding confidence intervals are computed by means of the `meta.MH` function for fixed effects meta-analysis as shown here

```
R> library("rmeta")
R> data("smoking", package = "HSAUR")
R> smokingOR <- meta.MH(smoking[["tt"]], smoking[["tc"]],
+     smoking[["qt"]], smoking[["qc"]],
+     names = rownames(smoking))
```

and the results can be inspected via a `summary` method – see Figure 12.1.

Before proceeding to the calculation of a combined effect size it will be helpful to graph the data by plotting confidence intervals for the odds ratios from each study (this is often known as a *forest plot* – see Sutton et al., 2000). The `plot` function applied to `smokingOR` produces such a plot, see Figure 12.2. It appears that the tendency in the trials considered was to favour nicotine gum but we need now to quantify this evidence in the form of an overall estimate of the odds ratio.

We shall use both the fixed effects and random effects approaches here so that we can compare results. For the fixed effects model (see Figure 12.1) the estimated overall log-odds ratio is 0.513 with a standard error of 0.066. This leads to an estimate of the overall odds ratio of 1.67, with a 95% confidence interval as given above. For the random effects model

```
R> smokingDSL <- meta.DSL(smoking[["tt"]], smoking[["tc"]],
+     smoking[["qt"]], smoking[["qc"]],
+     names = rownames(smoking))
R> print(smokingDSL)
```

```
Random effects ( DerSimonian-Laird ) meta-analysis
Call: meta.DSL(ntrt = smoking[["tt"]], nctrl =
    smoking[["tc"]], ptrt = smoking[["qt"]],
    pctrl = smoking[["qc"]], names = rownames(smoking))
Summary OR= 1.75    95% CI ( 1.48, 2.07 )
Estimated random effects variance: 0.05
```

the corresponding estimate is 1.751. Both models suggest that there is clear evidence that nicotine gum increases the odds of quitting. The random effects confidence interval is considerably wider than that from the fixed effects model; here the test of homogeneity of the studies is not significant implying that we might use the fixed effects results. But the test is not particularly powerful and it is more sensible to assume a priori that heterogeneity is present and so we use the results from the random effects model.

## 12.5 Meta-Regression

The examination of heterogeneity of the effect sizes from the studies in a meta-analysis begins with the formal test for its presence, although in most meta-analyses such heterogeneity can almost be assumed to be present. There

```
R> summary(smokingOR)

Fixed effects ( Mantel-Haenszel ) meta-analysis
Call: meta.MH(ntrt = smoking[["tt"]], nctrl = smoking[["tc"]],
    ptrt = smoking[["qt"]],
    pctrl = smoking[["qc"]], names = rownames(smoking))
------------------------------------
               OR (lower  95% upper)
Blondal89     1.85    0.99      3.46
Campbell91    0.98    0.50      1.92
Fagerstrom82  1.76    0.80      3.89
Fee82         1.53    0.77      3.05
Garcia89      2.95    1.01      8.62
Garvey00      2.49    1.43      4.34
Gross95       2.62    1.03      6.71
Hall85        2.03    0.78      5.29
Hall87        2.82    1.33      5.99
Hall96        0.87    0.46      1.64
Hjalmarson84  2.17    1.10      4.28
Huber88       6.00    2.57     14.01
Jarvis82      3.33    1.37      8.08
Jensen91      1.43    0.84      2.44
Killen84      1.33    0.43      4.15
Killen90      1.23    0.93      1.64
Malcolm80     3.52    0.85     14.54
McGovern92    1.17    0.70      1.94
Nakamura90    3.82    1.15     12.71
Niaura94      1.34    0.35      5.19
Pirie92       1.84    1.20      2.82
Puska79       1.46    0.78      2.75
Schneider85   1.71    0.52      5.62
Tonnesen88    2.12    0.93      4.86
Villa99       1.76    0.55      5.64
Zelman92      1.46    0.68      3.14
------------------------------------
Mantel-Haenszel OR =1.67 95% CI ( 1.47,1.9 )
Test for heterogeneity: X^2( 25 ) = 34.9 ( p-value 0.09 )
```

**Figure 12.1** R output of the `summary` method for `smokingOR`.

will be many possible sources of such heterogeneity and estimating how these various factors affect the observed effect sizes in the studies chosen is often of considerable interest and importance, indeed usually more important than the relatively simplistic use of meta-analysis to determine a single summary estimate of overall effect size. We can illustrate the process using the BCG vaccine data. We first find the estimate of the overall effect size from applying the fixed effects and the random effects models described previously:

```
R> plot(smokingOR, ylab = "")
```

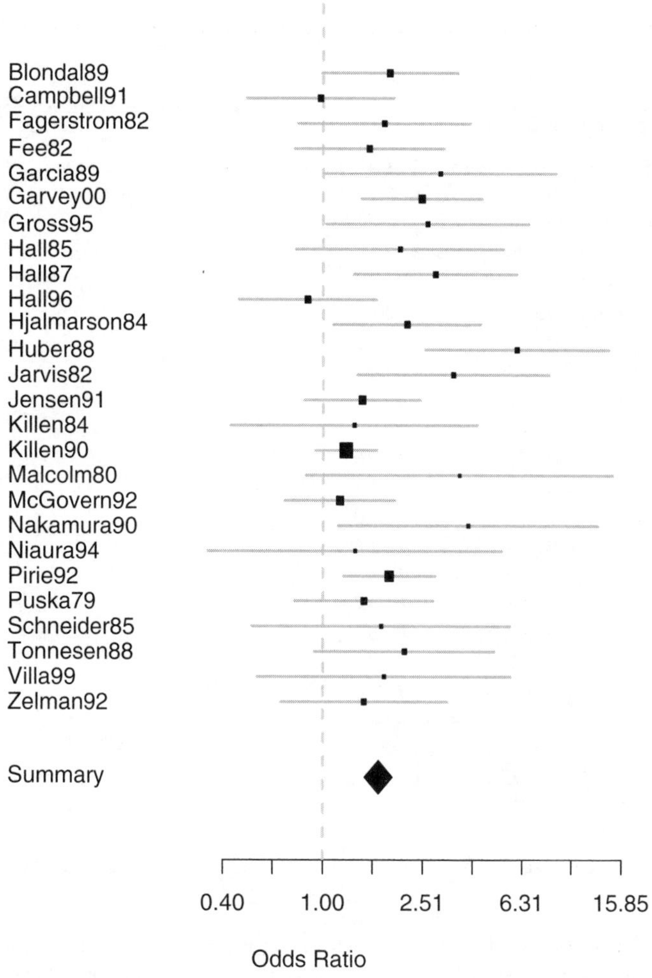


**Figure 12.2** Forest plot of observed effect sizes and 95% confidence intervals for the nicotine gum studies.

```
R> data("BCG", package = "HSAUR")
R> BCG_OR <- meta.MH(BCG[["BCGVacc"]], BCG[["NoVacc"]],
+     BCG[["BCGTB"]], BCG[["NoVaccTB"]], names = BCG$Study)
R> BCG_DSL <- meta.DSL(BCG[["BCGVacc"]], BCG[["NoVacc"]],
+     BCG[["BCGTB"]], BCG[["NoVaccTB"]], names = BCG$Study)
```

The results are inspected using the `summary` method as shown in Figures 12.3 and 12.4.

```
R> summary(BCG_OR)

Fixed effects ( Mantel-Haenszel ) meta-analysis
Call: meta.MH(ntrt = BCG[["BCGVacc"]], nctrl =
    BCG[["NoVacc"]], ptrt = BCG[["BCGTB"]],
    pctrl = BCG[["NoVaccTB"]], names = BCG$Study)
------------------------------------
      OR (lower  95% upper)
1  0.39     0.12          1.26
2  0.19     0.08          0.46
3  0.25     0.07          0.91
4  0.23     0.18          0.31
5  0.80     0.51          1.26
6  0.38     0.32          0.47
7  0.20     0.08          0.50
8  1.01     0.89          1.15
9  0.62     0.39          1.00
10 0.25     0.14          0.42
11 0.71     0.57          0.89
12 1.56     0.37          6.55
13 0.98     0.58          1.66
------------------------------------
Mantel-Haenszel OR =0.62 95% CI ( 0.57,0.68 )
Test for heterogeneity: X^2( 12 ) = 163.94 ( p-value 0 )
```

**Figure 12.3** R output of the `summary` method for `BCG_OR`.

For these data the test statistics for heterogeneity takes the value 163.16 which, with 12 degrees of freedom, is highly significant; there is strong evidence of heterogeneity in the 13 studies. Applying the random effects model to the data gives (see Figure 12.4) an estimated odds ratio 0.474, with a 95% confidence interval of $(0.325, 0.69)$ and an estimated between-study variance of 0.366.

To assess how the two covariates, latitude and year, relate to the observed effect sizes we shall use multiple linear regression but will weight each observation by $W_i = (\hat{\sigma}^2 + V_i^2)^{-1}, i = 1, \dots, 13$, where $\hat{\sigma}^2$ is the estimated between-study variance and $V_i^2$ is the estimated variance from the $i$th study. The required R code to fit the linear model via weighted least squares is:

```
R> summary(BCG_DSL)

Random effects ( DerSimonian-Laird ) meta-analysis
Call: meta.DSL(ntrt = BCG[["BCGVacc"]], nctrl =
    BCG[["NoVacc"]], ptrt = BCG[["BCGTB"]],
    pctrl = BCG[["NoVaccTB"]], names = BCG$Study)
------------------------------------
     OR (lower  95% upper)
1  0.39    0.12        1.26
2  0.19    0.08        0.46
3  0.25    0.07        0.91
4  0.23    0.18        0.31
5  0.80    0.51        1.26
6  0.38    0.32        0.47
7  0.20    0.08        0.50
8  1.01    0.89        1.15
9  0.62    0.39        1.00
10 0.25    0.14        0.42
11 0.71    0.57        0.89
12 1.56    0.37        6.55
13 0.98    0.58        1.66
------------------------------------
SummaryOR= 0.47  95% CI ( 0.32,0.69 )
Test for heterogeneity: X^2( 12 ) = 163.16 ( p-value 0 )
Estimated random effects variance: 0.37
```

**Figure 12.4** R output of the `summary` method for `BCG_DSL`.

```
R> studyweights <- 1/(BCG_DSL$tau2 + BCG_DSL$selogs)
R> y <- BCG_DSL$logs
R> BCG_mod <- lm(y ~ Latitude + Year, data = BCG,
+     weights = studyweights)
```

and the results of the `summary` method are shown in Figure 12.5. There is some evidence that latitude is associated with observed effect size, the log-odds ratio becoming increasingly negative as latitude increases, as we can see from a scatterplot of the two variables with the added weighted regression fit seen in Figure 12.6.

## 12.6 Publication Bias

The selection of studies to be integrated by a meta-analysis will clearly have a bearing on the conclusions reached. Selection is a matter of inclusion and exclusion and the judgements required are often difficult; Chalmers and Lau (1993) discuss the general issues involved, but here we shall concentrate on the particular potential problem of publication bias, which is a major problem, perhaps *the* major problem in meta-analysis.

```
R> summary(BCG_mod)

Call:
lm(formula = y ~ Latitude + Year, data = BCG, weights =
    studyweights)

Residuals:
     Min       1Q   Median       3Q      Max
-1.40868 -0.33622 -0.04847  0.25412  1.13362

Coefficients:
              Estimate Std. Error t value Pr(>|t|)
(Intercept) -14.521025  37.178382  -0.391   0.7043
Latitude     -0.026463   0.013553  -1.953   0.0794 .
Year          0.007442   0.018755   0.397   0.6998
---
Signif. codes:  0 '***' 0.001 '**' 0.01 '*' 0.05 '.' 0.1 ' ' 1

Residual standard error: 0.6862 on 10 degrees of freedom
Multiple R-Squared:  0.45,      Adjusted R-squared:  0.34
F-statistic: 4.091 on 2 and 10 DF,  p-value: 0.05033
```

**Figure 12.5** R output of the summary method for BCG_mod.

Ensuring that a meta-analysis is truly representative can be problematic. It has long been known that journal articles are not a representative sample of work addressed to any particular area of research (see Sterlin, 1959, Greenwald, 1975, Smith, 1980, for example). Research with statistically significant results is potentially more likely to be submitted and published than work with null or non-significant results (Easterbrook et al., 1991). The problem is made worse by the fact that many medical studies look at multiple outcomes, and there is a tendency for only those suggesting a significant effect to be mentioned when the study is written up. Outcomes which show no clear treatment effect are often ignored, and so will not be included in any later review of studies looking at those particular outcomes. Publication bias is likely to lead to an over-representation of positive results.

Clearly then it becomes of some importance to assess the likelihood of publication bias in any meta-analysis. A well-known, informal method of assessing publication bias is the so-called *funnel plot*. This assumes that the results from smaller studies will be more widely spread around the mean effect because of larger random error; a plot of a measure of the precision (such as inverse standard error or sample size) of the studies versus treatment effect from individual studies in a meta-analysis, should therefore be shaped like a funnel if there is no publication bias. If the chance of publication is greater for studies with statistically significant results, the shape of the funnel may become skewed.

```
R> plot(y ~ Latitude, data = BCG, ylab = "Estimated log-OR")
R> abline(lm(y ~ Latitude, data = BCG, weights = studyweights))
```

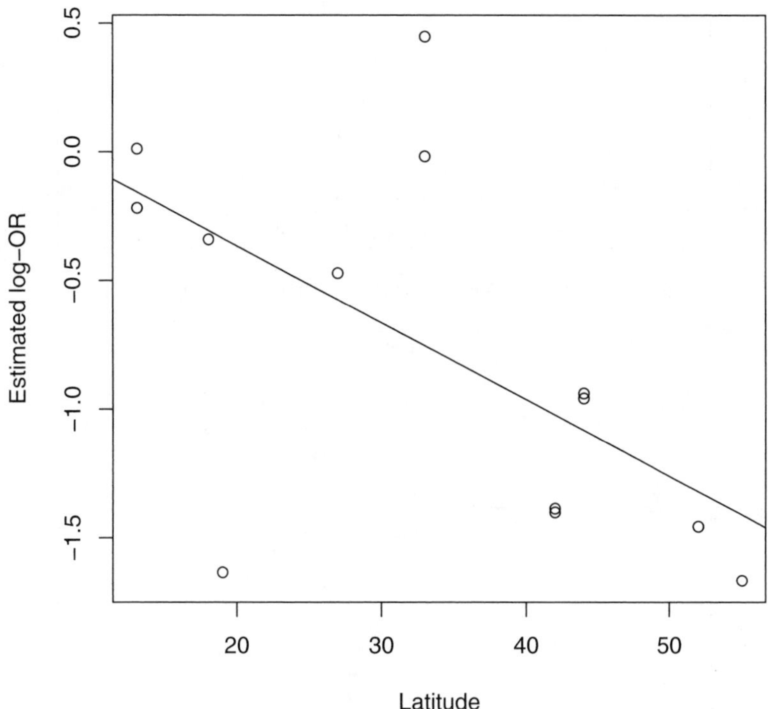


**Figure 12.6** Plot of observed effect size for the BCG vaccine data against latitude, with a weighted least squares regression fit shown in addition.

Example funnel plots, inspired by those shown in Duval and Tweedie (2000), are displayed in Figure 12.7. In the first of these plots, there is little evidence of publication bias, while in the second, there is definite asymmetry with a clear lack of studies in the bottom left hand corner of the plot.

We can construct a funnel plot for the nicotine gum data using the R code depicted with Figure 12.8. There does not appear to be any strong evidence of publication bias here.

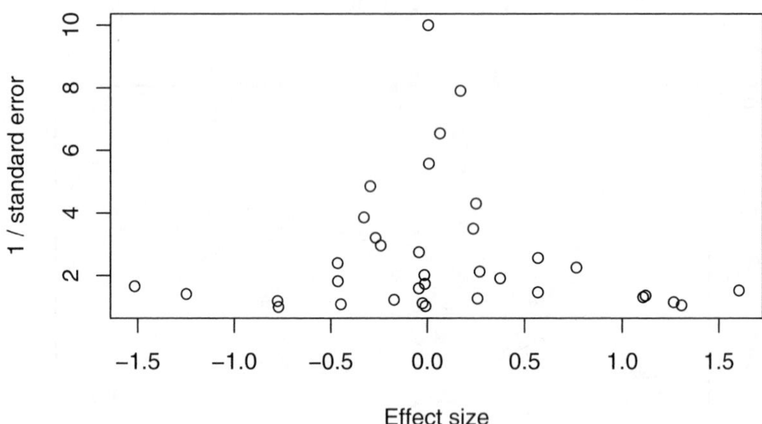


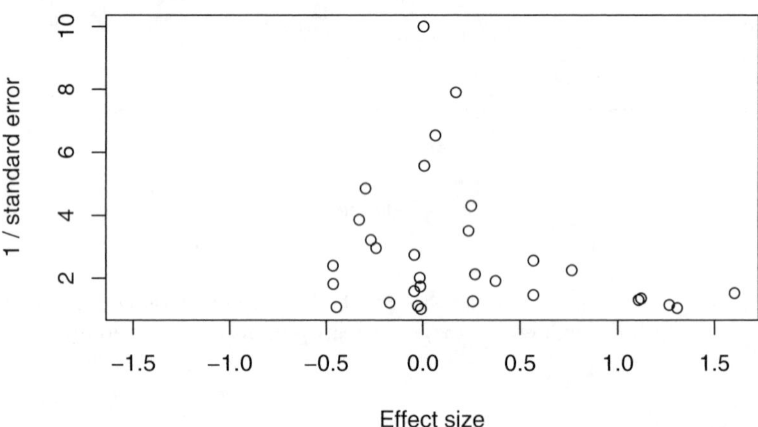


**Figure 12.7** Example funnel plots from simulated data. The asymmetry in the lower plot is a hint that a publication bias might be a problem.

```
R> funnelplot(smokingDSL$logs, smokingDSL$selogs,
+     summ = smokingDSL$logDSL,
+     xlim = c(-1.7, 1.7))
R> abline(v = 0, lty = 2)
```

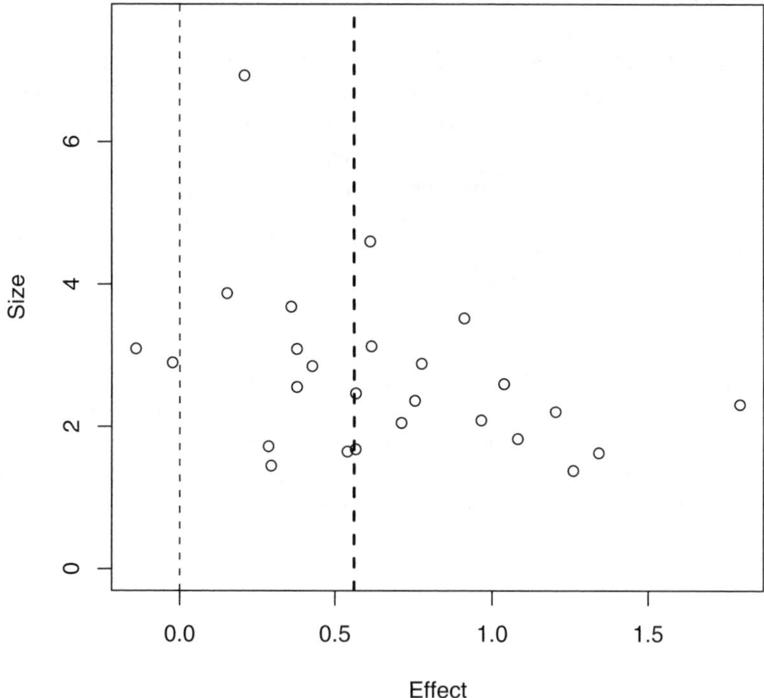


**Figure 12.8** Funnel plot for nicotine gum data.

## 12.7 Summary

It is probably fair to say that the majority of statisticians and clinicians are largely enthusiastic about the advantages of meta-analysis over the classical review, although there remain sceptics who feel that the conclusions from meta-analyses often go beyond what the techniques and the data justify. Some of their concerns are echoed in the following quotation from Oakes (1993);

> The term meta-analysis refers to the quantitative combination of data from independent trials. Where the results of such combination is a descriptive summary of the weight of the available evidence, the exercise is of undoubted value. Attempts to apply inferential methods, however, are subject to considerable methodologi-

cal and logical difficulties. The selection and quality of trials included, population bias and the specification of the population to which inference may properly be made are problems to which no satisfactory solutions have been proposed.

But despite such concerns the systematic review, in particular its quantitative component, meta-analysis, has had a major impact on medical science in the past ten years, and has been largely responsible for the development of evidence-based medical practice. One of the principal reasons that meta-analysis has been so successful is the large number of clinical trials that are now conducted. For example, the number of randomised clinical trials is now of the order of 10,000 per year. Synthesising results from many studies can be difficult, confusing and ultimately misleading. Meta-analysis has the potential to demonstrate treatment effects with a high degree of precision, possibly revealing small, but clinically important effects. But as with an individual clinical trial, careful planning, comprehensive data collection and a formal approach to statistical methods are necessary in order to achieve an acceptable and convincing meta-analysis.

## Exercises

Ex. 12.1 The data in Table 12.3 were collected for a meta-analysis of the effectiveness of Aspirin (versus placebo) in preventing death after a myocardial infarction (Fleiss, 1993). Calculate the log-odds ratio for each study and its variance, and then fit both a fixed effects and random effects model. Investigate the effect of possible publication bias.

**Table 12.3**: `aspirin` data. Meta-analysis on Aspirin and myocardial infarct, the table shows the number of deaths after placebo (`dp`), the total number subjects treated with placebo (`tp`) as well as the number of deaths after Aspirin (`da`) and the total number of subjects treated with Aspirin (`ta`).

| | `dp` | `tp` | `da` | `ta` |
|---|---|---|---|---|
| Elwood et al. (1974) | 67 | 624 | 49 | 615 |
| Coronary Drug Project Group (1976) | 64 | 77 | 44 | 757 |
| Elwood and Sweetman (1979) | 126 | 850 | 102 | 832 |
| Breddin et al. (1979) | 38 | 309 | 32 | 317 |
| Persantine-Aspirin Reinfarction Study Research Group (1980) | 52 | 406 | 85 | 810 |
| Aspirin Myocardial Infarction Study Research Group (1980) | 219 | 2257 | 346 | 2267 |
| ISIS-2 (Second International Study of Infarct Survival) Collaborative Group (1988) | 1720 | 8600 | 1570 | 8587 |

Ex. 12.2 The data in Table 12.4 shows the results of nine randomised trials comparing two different toothpastes for the prevention of caries development (see Everitt and Pickles, 2000). The outcomes in each trial was the change from baseline, in the decayed, missing (due to caries) and filled surface dental index (DMFS). Calculate an appropriate measure of effect size for each study and then carry out a meta-analysis of the results. What conclusions do you draw from the results?

**Table 12.4**: `toothpaste` data. Meta-analysis on trials comparing two toothpastes, the number of individuals in the study, the mean and the standard deviation for each study `A` and `B` are shown.

| Study | nA | meanA | sdA | nB | meanB | sdB |
|---|---|---|---|---|---|---|
| 1 | 134 | 5.96 | 4.24 | 113 | 4.72 | 4.72 |
| 2 | 175 | 4.74 | 4.64 | 151 | 5.07 | 5.38 |
| 3 | 137 | 2.04 | 2.59 | 140 | 2.51 | 3.22 |
| 4 | 184 | 2.70 | 2.32 | 179 | 3.20 | 2.46 |
| 5 | 174 | 6.09 | 4.86 | 169 | 5.81 | 5.14 |
| 6 | 754 | 4.72 | 5.33 | 736 | 4.76 | 5.29 |
| 7 | 209 | 10.10 | 8.10 | 209 | 10.90 | 7.90 |
| 8 | 1151 | 2.82 | 3.05 | 1122 | 3.01 | 3.32 |
| 9 | 679 | 3.88 | 4.85 | 673 | 4.37 | 5.37 |

Ex. 12.3 As an exercise in writing R code write your own meta-analysis function which allows the plotting of observed effect sizes and their associated confidence intervals (*forest plot*), estimates the overall effect size and its standard error by both the fixed effects and random effect models, and shows both on the constructed forest plot.

CHAPTER 13

# Principal Component Analysis: The Olympic Heptathlon

## 13.1 Introduction

The pentathlon for women was first held in Germany in 1928. Initially this consisted of the shot put, long jump, 100m, high jump and javelin events held over two days. In the 1964 Olympic Games the pentathlon became the first combined Olympic event for women, consisting now of the 80m hurdles, shot, high jump, long jump and 200m. In 1977 the 200m was replaced by the 800m and from 1981 the IAAF brought in the seven-event heptathlon in place of the pentathlon, with day one containing the events 100m hurdles, shot, high jump, 200m and day two, the long jump, javelin and 800m. A scoring system is used to assign points to the results from each event and the winner is the woman who accumulates the most points over the two days. The event made its first Olympic appearance in 1984.

In the 1988 Olympics held in Seoul, the heptathlon was won by one of the stars of women's athletics in the USA, Jackie Joyner-Kersee. The results for all 25 competitors in all seven disciplines are given in Table 13.1 (from Hand et al., 1994). We shall analyse these data using *principal component analysis* with a view to exploring the structure of the data and assessing how the derived principal component scores (see later) relate to the scores assigned by the official scoring system.

## 13.2 Principal Component Analysis

The basic aim of principal component analysis is to describe variation in a set of correlated variables, $x_1, x_2, \dots, x_q$, in terms of a new set of uncorrelated variables, $y_1, y_2, \dots, y_q$, each of which is a linear combination of the $x$ variables. The new variables are derived in decreasing order of 'importance' in the sense that $y_1$ accounts for as much of the variation in the original data amongst all linear combinations of $x_1, x_2, \dots, x_q$. Then $y_2$ is chosen to account for as much as possible of the remaining variation, subject to being uncorrelated with $y_1$ – and so on, i.e., forming an orthogonal coordinate system. The new variables defined by this process, $y_1, y_2, \dots, y_q$, are the principal components.

The general hope of principal component analysis is that the first few components will account for a substantial proportion of the variation in the original variables, $x_1, x_2, \dots, x_q$, and can, consequently, be used to provide a conve-

**Table 13.1**: `heptathlon` data. Results Olympic heptathlon, Seoul, 1988.

| | `hurdles` | `highjump` | `shot` | `run200m` | `longjump` | `javelin` | `run800m` | `score` |
|---|---|---|---|---|---|---|---|---|
| Joyner-Kersee (USA) | 12.69 | 1.86 | 15.80 | 22.56 | 7.27 | 45.66 | 128.51 | 7291 |
| John (GDR) | 12.85 | 1.80 | 16.23 | 23.65 | 6.71 | 42.56 | 126.12 | 6897 |
| Behmer (GDR) | 13.20 | 1.83 | 14.20 | 23.10 | 6.68 | 44.54 | 124.20 | 6858 |
| Sablovskaite (URS) | 13.61 | 1.80 | 15.23 | 23.92 | 6.25 | 42.78 | 132.24 | 6540 |
| Choubenkova (URS) | 13.51 | 1.74 | 14.76 | 23.93 | 6.32 | 47.46 | 127.90 | 6540 |
| Schulz (GDR) | 13.75 | 1.83 | 13.50 | 24.65 | 6.33 | 42.82 | 125.79 | 6411 |
| Fleming (AUS) | 13.38 | 1.80 | 12.88 | 23.59 | 6.37 | 40.28 | 132.54 | 6351 |
| Greiner (USA) | 13.55 | 1.80 | 14.13 | 24.48 | 6.47 | 38.00 | 133.65 | 6297 |
| Lajbnerova (CZE) | 13.63 | 1.83 | 14.28 | 24.86 | 6.11 | 42.20 | 136.05 | 6252 |
| Bouraga (URS) | 13.25 | 1.77 | 12.62 | 23.59 | 6.28 | 39.06 | 134.74 | 6252 |
| Wijnsma (HOL) | 13.75 | 1.86 | 13.01 | 25.03 | 6.34 | 37.86 | 131.49 | 6205 |
| Dimitrova (BUL) | 13.24 | 1.80 | 12.88 | 23.59 | 6.37 | 40.28 | 132.54 | 6171 |
| Scheider (SWI) | 13.85 | 1.86 | 11.58 | 24.87 | 6.05 | 47.50 | 134.93 | 6137 |
| Braun (FRG) | 13.71 | 1.83 | 13.16 | 24.78 | 6.12 | 44.58 | 142.82 | 6109 |
| Ruotsalainen (FIN) | 13.79 | 1.80 | 12.32 | 24.61 | 6.08 | 45.44 | 137.06 | 6101 |
| Yuping (CHN) | 13.93 | 1.86 | 14.21 | 25.00 | 6.40 | 38.60 | 146.67 | 6087 |
| Hagger (GB) | 13.47 | 1.80 | 12.75 | 25.47 | 6.34 | 35.76 | 138.48 | 5975 |
| Brown (USA) | 14.07 | 1.83 | 12.69 | 24.83 | 6.13 | 44.34 | 146.43 | 5972 |
| Mulliner (GB) | 14.39 | 1.71 | 12.68 | 24.92 | 6.10 | 37.76 | 138.02 | 5746 |
| Hautenauve (BEL) | 14.04 | 1.77 | 11.81 | 25.61 | 5.99 | 35.68 | 133.90 | 5734 |
| Kytola (FIN) | 14.31 | 1.77 | 11.66 | 25.69 | 5.75 | 39.48 | 133.35 | 5686 |
| Geremias (BRA) | 14.23 | 1.71 | 12.95 | 25.50 | 5.50 | 39.64 | 144.02 | 5508 |
| Hui-Ing (TAI) | 14.85 | 1.68 | 10.00 | 25.23 | 5.47 | 39.14 | 137.30 | 5290 |
| Jeong-Mi (KOR) | 14.53 | 1.71 | 10.83 | 26.61 | 5.50 | 39.26 | 139.17 | 5289 |
| Launa (PNG) | 16.42 | 1.50 | 11.78 | 26.16 | 4.88 | 46.38 | 163.43 | 4566 |

nient lower-dimensional summary of these variables that might prove useful for a variety of reasons.

In some applications, the principal components may be an end in themselves and might be amenable to interpretation in a similar fashion as the factors in an *exploratory factor analysis* (see Everitt and Dunn, 2001). More often they are obtained for use as a means of constructing a low-dimensional informative graphical representation of the data, or as input to some other analysis.

The low-dimensional representation produced by principal component analysis is such that

$$\sum_{r=1}^{n}\sum_{s=1}^{n}\left(d_{rs}^2 - \hat{d}_{rs}^2\right)$$

is minimised with respect to $\hat{d}_{rs}^2$. In this expression, $d_{rs}$ is the Euclidean distance (see Chapter 14) between observations $r$ and $s$ in the original $q$ dimensional space, and $\hat{d}_{rs}$ is the corresponding distance in the space of the first $m$ components.

As stated previously, the first principal component of the observations is that linear combination of the original variables whose sample variance is greatest amongst all possible such linear combinations. The second principal component is defined as that linear combination of the original variables that accounts for a maximal proportion of the remaining variance subject to being uncorrelated with the first principal component. Subsequent components are defined similarly. The question now arises as to how the coefficients specifying the linear combinations of the original variables defining each component are found? The algebra of *sample* principal components is summarised briefly.

The first principal component of the observations, $y_1$, is the linear combination

$$y_1 = a_{11}x_1 + a_{12}x_2 + \ldots, a_{1q}x_q$$

whose sample variance is greatest among all such linear combinations. Since the variance of $y_1$ could be increased without limit simply by increasing the coefficients $\mathbf{a}_1^\top = (a_{11}, a_{12}, \ldots, a_{1q})$ (here written in form of a vector for convenience), a restriction must be placed on these coefficients. As we shall see later, a sensible constraint is to require that the sum of squares of the coefficients, $\mathbf{a}_1^\top\mathbf{a}_1$, should take the value one, although other constraints are possible.

The second principal component $y_2 = \mathbf{a}_2^\top\mathbf{x}$ with $\mathbf{x} = (x_1, \ldots, x_q)$ is the linear combination with greatest variance subject to the two conditions $\mathbf{a}_2^\top\mathbf{a}_2 = 1$ and $\mathbf{a}_2^\top\mathbf{a}_1 = 0$. The second condition ensures that $y_1$ and $y_2$ are uncorrelated. Similarly, the $j$th principal component is that linear combination $y_j = \mathbf{a}_j^\top\mathbf{x}$ which has the greatest variance subject to the conditions $\mathbf{a}_j^\top\mathbf{a}_j = 1$ and $\mathbf{a}_j^\top\mathbf{a}_i = 0$ for $(i < j)$.

To find the coefficients defining the first principal component we need to choose the elements of the vector $\mathbf{a}_1$ so as to maximise the variance of $y_1$ subject to the constraint $\mathbf{a}_1^\top\mathbf{a}_1 = 1$.

To maximise a function of several variables subject to one or more constraints, the method of *Lagrange multipliers* is used. In this case this leads to the solution that $\mathbf{a}_1$ is the eigenvector of the sample covariance matrix, $\mathbf{S}$, corresponding to its largest eigenvalue – full details are given in Morrison (2005).

The other components are derived in similar fashion, with $\mathbf{a}_j$ being the eigenvector of $\mathbf{S}$ associated with its $j$th largest eigenvalue. If the eigenvalues of $\mathbf{S}$ are $\lambda_1, \lambda_2, \dots, \lambda_q$, then since $\mathbf{a}_j^\top \mathbf{a}_j = 1$, the variance of the $j$th component is given by $\lambda_j$.

The total variance of the $q$ principal components will equal the total variance of the original variables so that

$$\sum_{j=1}^{q} \lambda_j = s_1^2 + s_2^2 + \dots + s_q^2$$

where $s_j^2$ is the sample variance of $x_j$. We can write this more concisely as

$$\sum_{j=1}^{q} \lambda_j = \text{trace}(\mathbf{S}).$$

Consequently, the $j$th principal component accounts for a proportion $P_j$ of the total variation of the original data, where

$$P_j = \frac{\lambda_j}{\text{trace}(\mathbf{S})}.$$

The first $m$ principal components, where $m < q$, account for a proportion

$$P^{(m)} = \frac{\sum_{j=1}^{m} \lambda_j}{\text{trace}(\mathbf{S})}.$$

## 13.3 Analysis Using R

To begin it will help to score all the seven events in the same direction, so that 'large' values are 'good'. We will recode the running events to achieve this;

```
R> data("heptathlon", package = "HSAUR")
R> heptathlon$hurdles <- max(heptathlon$hurdles) -
+      heptathlon$hurdles
R> heptathlon$run200m <- max(heptathlon$run200m) -
+      heptathlon$run200m
R> heptathlon$run800m <- max(heptathlon$run800m) -
+      heptathlon$run800m
```

Figure 13.1 shows a scatterplot matrix of the results from the 25 competitors on the seven events. We see that most pairs of events are positively correlated to a greater or lesser degree. The exceptions all involve the javelin event – this is the only really 'technical' event and it is clear that training to become

```
R> score <- which(colnames(heptathlon) == "score")
R> plot(heptathlon[, -score])
```

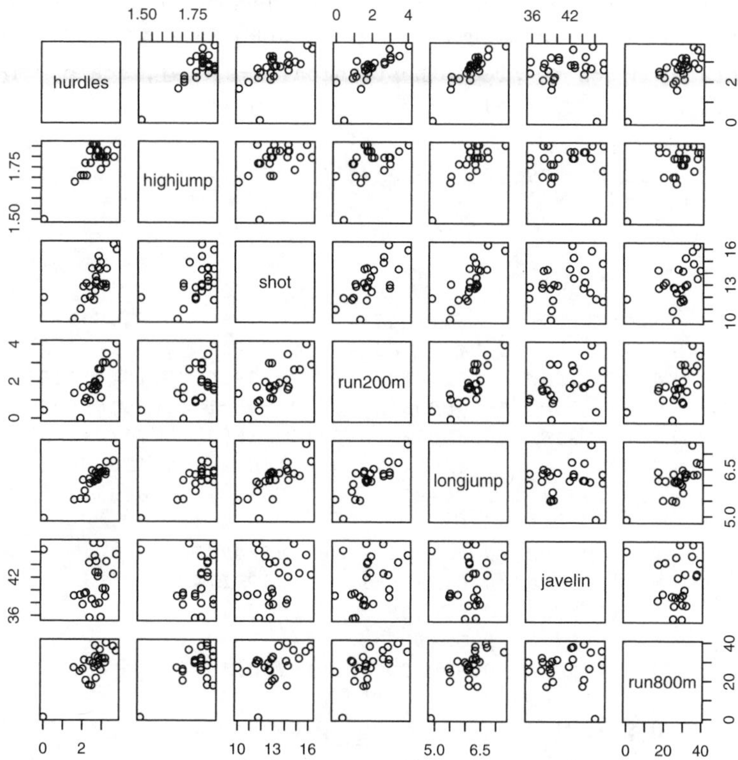


**Figure 13.1** Scatterplot matrix for the heptathlon data.

successful in the other six 'power'-based events makes this event difficult for the majority of the competitors. We can examine the numerical values of the correlations by applying the cor function

```
R> round(cor(heptathlon[, -score]), 2)

         hurdles highjump shot run200m longjump javelin run800m
hurdles     1.00     0.81 0.65    0.77     0.91    0.01    0.78
highjump    0.81     1.00 0.44    0.49     0.78    0.00    0.59
shot        0.65     0.44 1.00    0.68     0.74    0.27    0.42
run200m     0.77     0.49 0.68    1.00     0.82    0.33    0.62
longjump    0.91     0.78 0.74    0.82     1.00    0.07    0.70
javelin     0.01     0.00 0.27    0.33     0.07    1.00   -0.02
run800m     0.78     0.59 0.42    0.62     0.70   -0.02    1.00
```

This correlation matrix demonstrates again the points made earlier.

A principal component analysis of the data can be applied using the `prcomp` function. The result is a list containing the coefficients defining each component (sometimes referred to as *loadings*), the principal component scores, etc. The required code is (omitting the `score` variable)

```
R> heptathlon_pca <- prcomp(heptathlon[, -score], scale = TRUE)
R> print(heptathlon_pca)
```

```
Standard deviations:
[1] 2.1119364 1.0928497 0.7218131 0.6761411 0.4952441 0.2701029
[7] 0.2213617

Rotation:
                PC1         PC2         PC3         PC4
hurdles  -0.4528710  0.15792058 -0.04514996  0.02653873
highjump -0.3771992  0.24807386 -0.36777902  0.67999172
shot     -0.3630725 -0.28940743  0.67618919  0.12431725
run200m  -0.4078950 -0.26038545  0.08359211 -0.36106580
longjump -0.4562318  0.05587394  0.13931653  0.11129249
javelin  -0.0754090 -0.84169212 -0.47156016  0.12079924
run800m  -0.3749594  0.22448984 -0.39585671 -0.60341130
                 PC5         PC6         PC7
hurdles  -0.09494792 -0.78334101  0.38024707
highjump  0.01879888  0.09939981 -0.43393114
shot      0.51165201 -0.05085983 -0.21762491
run200m  -0.64983404  0.02495639 -0.45338483
longjump -0.18429810  0.59020972  0.61206388
javelin   0.13510669 -0.02724076  0.17294667
run800m   0.50432116  0.15555520 -0.09830963
```

The `summary` method can be used for further inspection of the details:

```
R> summary(heptathlon_pca)
```

```
Importance of components:
                         PC1   PC2    PC3    PC4    PC5    PC6
Standard deviation     2.112 1.093 0.7218 0.6761 0.4952 0.2701
Proportion of Variance 0.637 0.171 0.0744 0.0653 0.0350 0.0104
Cumulative Proportion  0.637 0.808 0.8822 0.9475 0.9826 0.9930
                         PC7
Standard deviation     0.221
Proportion of Variance 0.007
Cumulative Proportion  1.000
```

The linear combination for the first principal component is

```
R> a1 <- heptathlon_pca$rotation[, 1]
R> a1
```

```
   hurdles   highjump       shot    run200m   longjump
-0.4528710 -0.3771992 -0.3630725 -0.4078950 -0.4562318
   javelin    run800m
-0.0754090 -0.3749594
```

We see that the 200m and long jump competitions receive the highest weight but the javelin result is less important. For computing the first principal component, the data need to be rescaled appropriately. The center and the scaling used by `prcomp` internally can be extracted from the `heptathlon_pca` via

```
R> center <- heptathlon_pca$center
R> scale <- heptathlon_pca$scale
```

Now, we can apply the `scale` function to the data and multiply with the loadings matrix in order to compute the first principal component score for each competitor

```
R> hm <- as.matrix(heptathlon[, -score])
R> drop(scale(hm, center = center, scale = scale) %*%
+      heptathlon_pca$rotation[, 1])
```

```
Joyner-Kersee (USA)          John (GDR)          Behmer (GDR)
       -4.121447626        -2.882185935          -2.649633766
 Sablovskaite (URS)  Choubenkova (URS)          Schulz (GDR)
       -1.343351210        -1.359025696          -1.043847471
      Fleming (AUS)       Greiner (USA)      Lajbnerova (CZE)
       -1.100385639        -0.923173639          -0.530250689
      Bouraga (URS)       Wijnsma (HOL)       Dimitrova (BUL)
       -0.759819024        -0.556268302          -1.186453832
     Scheider (SWI)         Braun (FRG)   Ruotsalainen (FIN)
        0.015461226         0.003774223           0.090747709
       Yuping (CHN)         Hagger (GB)           Brown (USA)
       -0.137225440         0.171128651           0.519252646
       Mulliner (GB)   Hautenauve (BEL)          Kytola (FIN)
        1.125481833         1.085697646           1.447055499
     Geremias (BRA)      Hui-Ing (TAI)        Jeong-Mi (KOR)
        2.014029620         2.880298635           2.970118607
        Launa (PNG)
        6.270021972
```

or, more conveniently, by extracting the first from all precomputed principal components

```
R> predict(heptathlon_pca)[, 1]
```

```
Joyner-Kersee (USA)          John (GDR)          Behmer (GDR)
       -4.121447626        -2.882185935          -2.649633766
 Sablovskaite (URS)  Choubenkova (URS)          Schulz (GDR)
       -1.343351210        -1.359025696          -1.043847471
      Fleming (AUS)       Greiner (USA)      Lajbnerova (CZE)
       -1.100385639        -0.923173639          -0.530250689
      Bouraga (URS)       Wijnsma (HOL)       Dimitrova (BUL)
       -0.759819024        -0.556268302          -1.186453832
     Scheider (SWI)         Braun (FRG)   Ruotsalainen (FIN)
        0.015461226         0.003774223           0.090747709
       Yuping (CHN)         Hagger (GB)           Brown (USA)
       -0.137225440         0.171128651           0.519252646
       Mulliner (GB)   Hautenauve (BEL)          Kytola (FIN)
```

```
R> plot(heptathlon_pca)
```

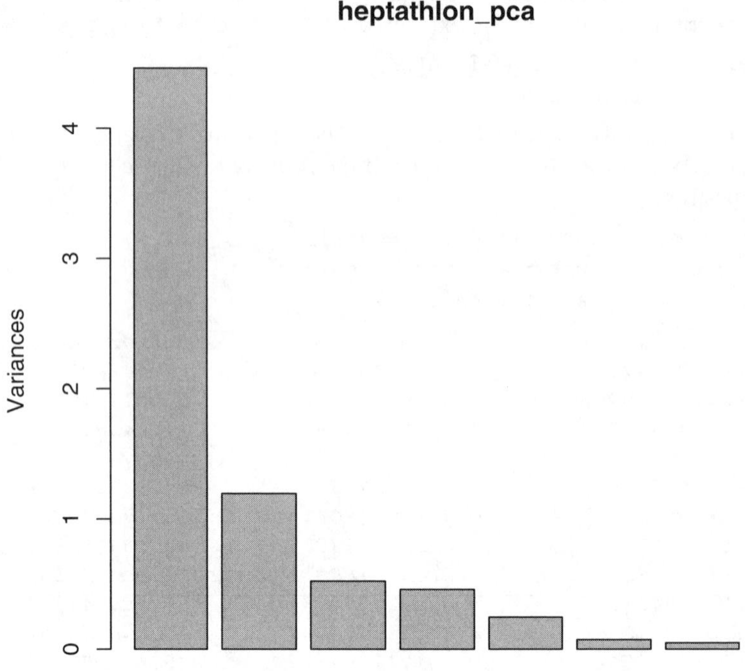


**Figure 13.2** Barplot of the variances explained by the principal components.

```
     1.125481833         1.085697646          1.447055499
 Geremias (BRA)       Hui-Ing (TAI)      Jeong-Mi (KOR)
     2.014029620         2.880298635          2.970118607
     Launa (PNG)
     6.270021972
```

The first two components account for 81% of the variance. A barplot of each component's variance (see Figure 13.2) shows how the first two components dominate. A plot of the data in the space of the first two principal components, with the points labelled by the name of the corresponding competitor can be produced as shown with Figure 13.3. In addition, the first two loadings for the events are given in a second coordinate system, also illustrating the special role of the javelin event. This graphical representation is known as *biplot* (Gabriel, 1971).

```
R> biplot(heptathlon_pca, col = c("gray", "black"))
```

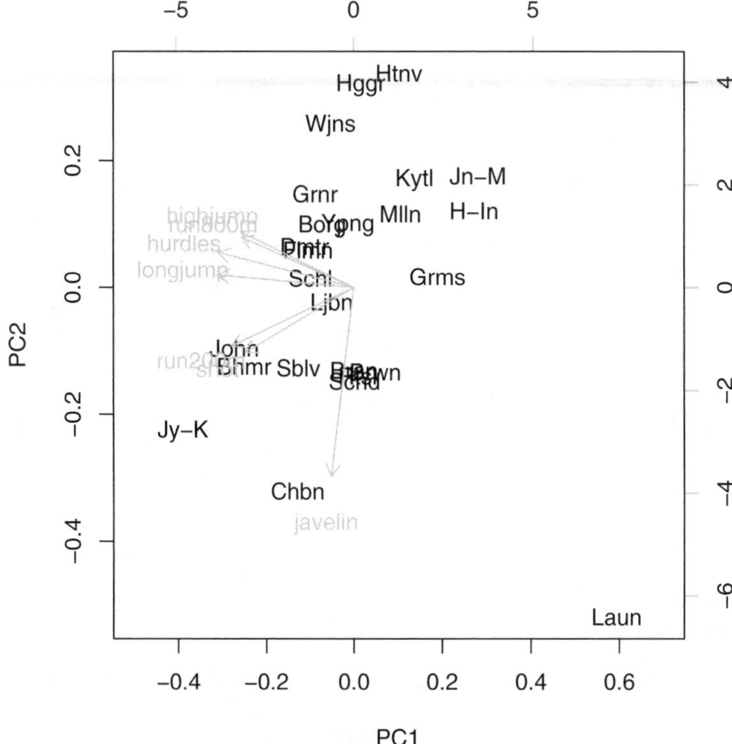


**Figure 13.3** Biplot of the (scaled) first two principal components.

The correlation between the score given to each athlete by the standard scoring system used for the heptathlon and the first principal component score can be found from

```
R> cor(heptathlon$score, heptathlon_pca$x[, 1])
[1] -0.9910978
```

This implies that the first principal component is in good agreement with the score assigned to the athletes by official Olympic rules; a scatterplot of the official score and the first principal component is given in Figure 13.4.

## 13.4 Summary

Principal components look for a few linear combinations of the original variables that can be used to summarise a data set, losing in the process as little

```
R> plot(heptathlon$score, heptathlon_pca$x[, 1])
```

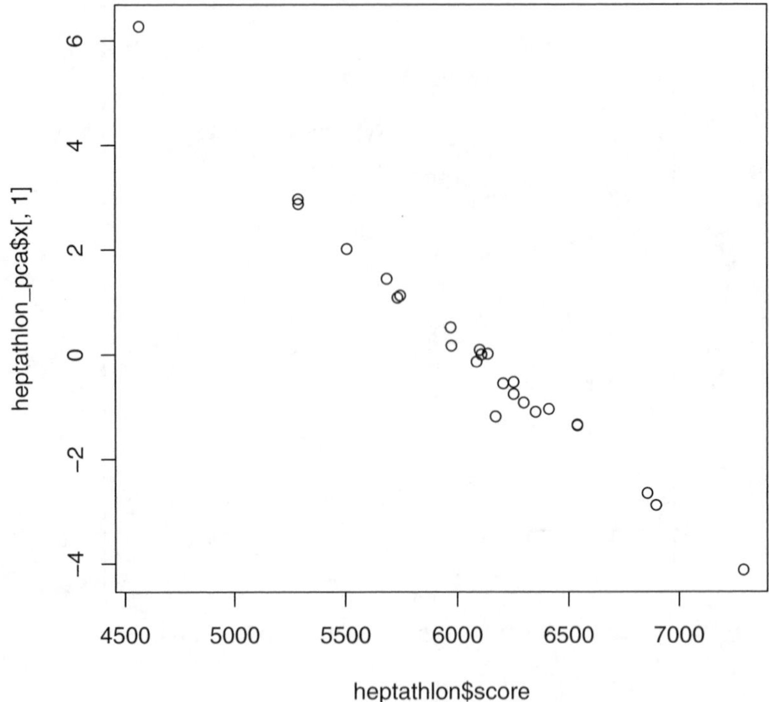


**Figure 13.4** Scatterplot of the score assigned to each athlete in 1988 and the first principal component.

information as possible. The derived variables might be used in a variety of ways, in particular for simplifying later analyses and providing informative plots of the data. The method consists of transforming a set of correlated variables to a new set of variables which are uncorrelated. Consequently it should be noted that if the original variables are themselves almost uncorrelated there is little point in carrying out a principal components analysis, since it will merely find components which are close to the original variables but arranged in decreasing order of variance.

**Exercises**

Ex. 13.1 Apply principal components analysis to the covariance matrix of the heptathlon data (excluding the score variable) and compare your results with those given in the text, derived from the correlation matrix of the data. Which results do you think are more appropriate for these data?

Ex. 13.2 The data in Table 13.2 below give measurements on five meteorological variables over an 11-year period (taken from Everitt and Dunn, 2001). The variables are;

`year`: the corresponding year,
`rainNovDec`: rainfall in November and December (mm),
`temp`: average July temperature,
`rainJuly`: rainfall in July (mm),
`radiation`: radiation in July (curies), and
`yield`: average harvest yield (quintals per hectare).

Carry out a principal components analysis of both the covariance matrix and the correlation matrix of the data and compare the results. Which set of components leads to the most meaningful interpretation?

**Table 13.2**: `meteo` data. Meteorological measurements in an 11-year period.

| year | rainNovDec | temp | rainJuly | radiation | yield |
|---|---|---|---|---|---|
| 1920-21 | 87.9 | 19.6 | 1.0 | 1661 | 28.37 |
| 1921-22 | 89.9 | 15.2 | 90.1 | 968 | 23.77 |
| 1922-23 | 153.0 | 19.7 | 56.6 | 1353 | 26.04 |
| 1923-24 | 132.1 | 17.0 | 91.0 | 1293 | 25.74 |
| 1924-25 | 88.8 | 18.3 | 93.7 | 1153 | 26.68 |
| 1925-26 | 220.9 | 17.8 | 106.9 | 1286 | 24.29 |
| 1926-27 | 117.7 | 17.8 | 65.5 | 1104 | 28.00 |
| 1927-28 | 109.0 | 18.3 | 41.8 | 1574 | 28.37 |
| 1928-29 | 156.1 | 17.8 | 57.4 | 1222 | 24.96 |
| 1929-30 | 181.5 | 16.8 | 140.6 | 902 | 21.66 |
| 1930-31 | 181.4 | 17.0 | 74.3 | 1150 | 24.37 |

*Source*: From Everitt, B. S. and Dunn, G., *Applied Multivariate Data Analysis, 2nd Edition*, Arnold, London, 2001. With permission.

Ex. 13.3 The correlations below are for the calculus measurements for the six anterior mandibular teeth. Find all six principal components of the data and use a screeplot to suggest how many components are needed to adequately account for the observed correlations. Can you interpret the components?

**Table 13.3**: Correlations for calculus measurements for the six anterior mandibular teeth.

| | | | | | |
|---|---|---|---|---|---|
| 1.00 | | | | | |
| 0.54 | 1.00 | | | | |
| 0.34 | 0.65 | 1.00 | | | |
| 0.37 | 0.65 | 0.84 | 1.00 | | |
| 0.36 | 0.59 | 0.67 | 0.80 | 1.00 | |
| 0.62 | 0.49 | 0.43 | 0.42 | 0.55 | 1.00 |

CHAPTER 14

# Multidimensional Scaling: British Water Voles and Voting in US Congress

## 14.1 Introduction

Corbet et al. (1970) report a study of water voles (genus *Arvicola*) in which the aim was to compare British populations of these animals with those in Europe, to investigate whether more than one species might be present in Britain. The original data consisted of observations of the presence or absence of 13 characteristics in about 300 water vole skulls arising from six British populations and eight populations from the rest of Europe. Table 14.1 gives a distance matrix derived from the data as described in Corbet et al. (1970).

Romesburg (1984) gives a set of data that shows the number of times 15 congressmen from New Jersey voted differently in the House of Representatives on 19 environmental bills. Abstentions are not recorded, but two congressmen abstained more frequently than the others, these being Sandman (nine abstentions) and Thompson (six abstentions). The data are available in Table 14.2 and one question of interest is can party affiliations be detected?

## 14.2 Multidimensional Scaling

The data in Tables 14.1 and 14.2 are both examples of *proximity matrices*. The elements of such matrices attempt to quantify how similar are stimuli, objects, individuals, etc. In Table 14.1 the values measure the 'distance' between populations of water voles; in Table 14.2 it is the similarity of the voting behaviour of the congressmen that is measured. Models are fitted to proximities in order to clarify, display and possibly explain any structure or pattern not readily apparent in the collection of numerical values. In some areas, particularly psychology, the ultimate goal in the analysis of a set of proximities is more specifically theories for explaining similarity judgements, or in other words, finding an answer to the question 'what makes things seem alike or seem different?' Here though we will concentrate on how proximity data can be best displayed to aid in uncovering any interesting structure.

The class of techniques we shall consider here, generally collected under the label *multidimensional scaling* (MDS), has the unifying feature that they seek to represent an observed proximity matrix by a simple geometrical model or map. Such a model consists of a series of say $q$-dimensional coordinate values,

**Table 14.1**: `watervoles` data. Water voles data – dissimilarity matrix.

| | Srry | Shrp | Yrks | Prth | Abrd | ElnG | Alps | Ygsl | Grmn | Nrwy | PyrI | PyII | NrtS | SthS |
|---|---|---|---|---|---|---|---|---|---|---|---|---|---|---|
| Surrey | 0.000 | | | | | | | | | | | | | |
| Shropshire | 0.099 | 0.000 | | | | | | | | | | | | |
| Yorkshire | 0.033 | 0.022 | 0.000 | | | | | | | | | | | |
| Perthshire | 0.183 | 0.114 | 0.042 | 0.000 | | | | | | | | | | |
| Aberdeen | 0.148 | 0.224 | 0.059 | 0.068 | 0.000 | | | | | | | | | |
| Elean Gamhna | 0.198 | 0.039 | 0.053 | 0.085 | 0.051 | 0.000 | | | | | | | | |
| Alps | 0.462 | 0.266 | 0.322 | 0.435 | 0.268 | 0.025 | 0.000 | | | | | | | |
| Yugoslavia | 0.628 | 0.442 | 0.444 | 0.406 | 0.240 | 0.129 | 0.014 | 0.000 | | | | | | |
| Germany | 0.113 | 0.070 | 0.046 | 0.047 | 0.034 | 0.002 | 0.106 | 0.129 | 0.000 | | | | | |
| Norway | 0.173 | 0.119 | 0.162 | 0.331 | 0.177 | 0.039 | 0.089 | 0.237 | 0.071 | 0.000 | | | | |
| Pyrenees I | 0.434 | 0.419 | 0.339 | 0.505 | 0.469 | 0.390 | 0.315 | 0.349 | 0.151 | 0.430 | 0.000 | | | |
| Pyrenees II | 0.762 | 0.633 | 0.781 | 0.700 | 0.758 | 0.625 | 0.469 | 0.618 | 0.440 | 0.538 | 0.607 | 0.000 | | |
| North Spain | 0.530 | 0.389 | 0.482 | 0.579 | 0.597 | 0.498 | 0.374 | 0.562 | 0.247 | 0.383 | 0.387 | 0.084 | 0.000 | |
| South Spain | 0.586 | 0.435 | 0.550 | 0.530 | 0.552 | 0.509 | 0.369 | 0.471 | 0.234 | 0.346 | 0.456 | 0.090 | 0.038 | 0.000 |

**Table 14.2:** `voting` data. House of Representatives voting data.

| | Hnt | Snd | Hwr | Thm | Fry | Frs | Wdn | Roe | Hlt | Rdn | Mns | Rnl | Mrz | Dnl | Ptt |
|---|---|---|---|---|---|---|---|---|---|---|---|---|---|---|---|
| Hunt(R) | 0 | | | | | | | | | | | | | | |
| Sandman(R) | 8 | 0 | | | | | | | | | | | | | |
| Howard(D) | 15 | 17 | 0 | | | | | | | | | | | | |
| Thompson(D) | 15 | 12 | 9 | 0 | | | | | | | | | | | |
| Freylinghuysen(R) | 10 | 13 | 16 | 14 | 0 | | | | | | | | | | |
| Forsythe(R) | 9 | 13 | 12 | 12 | 8 | 0 | | | | | | | | | |
| Widnall(R) | 7 | 12 | 15 | 13 | 9 | 7 | 0 | | | | | | | | |
| Roe(D) | 15 | 16 | 5 | 10 | 13 | 12 | 17 | 0 | | | | | | | |
| Heltoski(D) | 16 | 17 | 5 | 8 | 14 | 11 | 16 | 4 | 0 | | | | | | |
| Rodino(D) | 14 | 15 | 6 | 8 | 12 | 10 | 15 | 5 | 3 | 0 | | | | | |
| Minish(D) | 15 | 16 | 5 | 8 | 12 | 9 | 14 | 5 | 2 | 1 | 0 | | | | |
| Rinaldo(R) | 16 | 17 | 4 | 6 | 12 | 10 | 15 | 3 | 1 | 2 | 1 | 0 | | | |
| Maraziti(R) | 7 | 13 | 11 | 15 | 10 | 6 | 10 | 12 | 13 | 11 | 12 | 12 | 0 | | |
| Daniels(D) | 11 | 12 | 10 | 10 | 11 | 6 | 11 | 7 | 7 | 4 | 5 | 6 | 9 | 0 | |
| Patten(D) | 13 | 16 | 7 | 7 | 11 | 10 | 13 | 6 | 5 | 6 | 5 | 4 | 13 | 9 | 0 |

$n$ in number, where $n$ is the number of rows (and columns) of the proximity matrix, and an associated measure of distance between pairs of points. Each point is used to represent one of the stimuli in the resulting spatial model for the proximities and the objective of a multidimensional approach is to determine both the dimensionality of the model (i.e., the value of $q$) that provides an adequate 'fit', and the positions of the points in the resulting $q$-dimensional space. Fit is judged by some numerical index of the correspondence between the observed proximities and the inter-point distances. In simple terms this means that the larger the perceived distance or dissimilarity between two stimuli (or the smaller their similarity), the further apart should be the points representing them in the final geometrical model.

A number of inter-point distance measures might be used, but by far the most common is *Euclidean distance.* For two points, $i$ and $j$, with $q$-dimensional coordinate values, $\mathbf{x}_i = (x_{i1}, x_{i2}, \ldots, x_{iq})$ and $\mathbf{x}_j = (x_{j1}, x_{j2}, \ldots, x_{jq})$ the Euclidean distance is defined as

$$d_{ij} = \sqrt{\sum_{k=1}^{q} (x_{ik} - x_{jk})^2}.$$

Having decided on a suitable distance measure the problem now becomes one of estimating the coordinate values to represent the stimuli, and this is achieved by optimizing the chosen goodness of fit index measuring how well the fitted distances match the observed proximities. A variety of optimisation schemes combined with a variety of goodness of fit indices leads to a variety of MDS methods. For details see, for example, Everitt and Rabe-Hesketh (1997). Here we give a brief account of two methods, *classical scaling*, and *non-metric scaling* which will then be used to analyse the two data sets described earlier.

### *14.2.1 Classical Multidimensional Scaling*

Classical scaling provides one answer to how we estimate $q$, and the $n$, $q$-dimensional, coordinate values $\mathbf{x}_1, \mathbf{x}_2, \ldots, \mathbf{x}_n$, from the observed proximity matrix, based on the work of Young and Householder (1938). To begin we must note that there is no unique set of coordinate values since the Euclidean distances involved are unchanged by shifting the whole configuration of points from one place to another, or by rotation or reflection of the configuration. In other words, we cannot uniquely determine either the location or the orientation of the configuration. The location problem is usually overcome by placing the mean vector of the configuration at the origin. The orientation problem means that any configuration derived can be subjected to an arbitrary *orthogonal transformation.* Such transformations can often be used to facilitate the interpretation of solutions as will be seen later.

To begin our account of the method we shall assume that the proximity matrix we are dealing with is a matrix of Euclidean distances $\mathbf{D}$ derived from a raw data matrix, $\mathbf{X}$. Previously we saw how to calculate Euclidean distances from $\mathbf{X}$; multidimensional scaling is essentially concerned with the reverse problem, given the distances how do we find $\mathbf{X}$?

An $n \times n$ inner products matrix $\mathbf{B}$ is first calculated as $\mathbf{B} = \mathbf{X}\mathbf{X}^\top$, the elements of $\mathbf{B}$ are given by

$$b_{ij} = \sum_{k=1}^{q} x_{ik}x_{jk}. \tag{14.1}$$

It is easy to see that the squared Euclidean distances between the rows of $\mathbf{X}$ can be written in terms of the elements of $\mathbf{B}$ as

$$d_{ij}^2 = b_{ii} + b_{jj} - 2b_{ij}. \tag{14.2}$$

If the $b$s could be found in terms of the $d$s as in the equation above, then the required coordinate value could be derived by factoring $\mathbf{B} = \mathbf{X}\mathbf{X}^\top$.

No unique solution exists unless a location constraint is introduced; usually the centre of the points $\bar{\mathbf{x}}$ is set at the origin, so that $\sum_{i=1}^{n} x_{ik} = 0$ for all $k$.

These constraints and the relationship given in (14.1) imply that the sum of the terms in any row of $\mathbf{B}$ must be zero.

Consequently, summing the relationship given in (14.2) over $i$, over $j$ and finally over both $i$ and $j$, leads to the following series of equations:

$$\begin{aligned}
\sum_{i=1}^{n} d_{ij}^2 &= \text{trace}(\mathbf{B}) + nb_{jj} \\
\sum_{j=1}^{n} d_{ij}^2 &= \text{trace}(\mathbf{B}) + nb_{ii} \\
\sum_{i=1}^{n}\sum_{j=1}^{n} d_{ij}^2 &= 2n \times \text{trace}(\mathbf{B})
\end{aligned}$$

where trace($\mathbf{B}$) is the trace of the matrix $\mathbf{B}$. The elements of $\mathbf{B}$ can now be found in terms of squared Euclidean distances as

$$b_{ij} = -\frac{1}{2}\left(d_{ij}^2 - n^{-1}\sum_{j=1}^{n} d_{ij}^2 - n^{-1}\sum_{i=1}^{n} d_{ij}^2 - n^{-2}\sum_{i=1}^{n}\sum_{j=1}^{n} d_{ij}^2\right).$$

Having now derived the elements of $\mathbf{B}$ in terms of Euclidean distances, it remains to factor it to give the coordinate values. In terms of its singular value decomposition $\mathbf{B}$ can be written as

$$\mathbf{B} = \mathbf{V}\Lambda\mathbf{V}^\top$$

where $\Lambda = \text{diag}(\lambda_1, \dots, \lambda_n)$ is the diagonal matrix of eigenvalues of $\mathbf{B}$ and $\mathbf{V}$ the corresponding matrix of eigenvectors, normalised so that the sum of squares of their elements is unity, that is, $\mathbf{V}^\top\mathbf{V} = \mathbf{I}_n$. The eigenvalues are assumed labeled such that $\lambda_1 \geq \lambda_2 \geq \dots \geq \lambda_n$.

When the matrix of Euclidian distances $\mathbf{D}$ arises from an $n \times k$ matrix of full column rank, then the rank of $\mathbf{B}$ is $k$, so that the last $n - k$ of its eigenvalues will be zero. So $\mathbf{B}$ can be written as $\mathbf{B} = \mathbf{V}_1\Lambda_1\mathbf{V}_1^\top$, where $\mathbf{V}_1$ contains the first $k$ eigenvectors and $\Lambda_1$ the $q$ non-zero eigenvalues. The required coordinate values are thus $\mathbf{X} = \mathbf{V}_1\Lambda_1^{1/2}$, where $\Lambda_1^{1/2} = \text{diag}(\sqrt{\lambda_1}, \ldots, \sqrt{\lambda_k})$.

The best fitting $k$-dimensional representation is given by the $k$ eigenvectors of $\mathbf{B}$ corresponding to the $k$ largest eigenvalues. The adequacy of the $k$-dimensional representation can be judged by the size of the criterion

$$P_k = \frac{\sum_{i=1}^{k} \lambda_i}{\sum_{i=1}^{n-1} \lambda_i}.$$

Values of $P_k$ of the order of 0.8 suggest a reasonable fit.

When the observed dissimilarity matrix is not Euclidean, the matrix $\mathbf{B}$ is not positive-definite. In such cases some of the eigenvalues of $\mathbf{B}$ will be negative; corresponding, some coordinate values will be complex numbers. If, however, $\mathbf{B}$ has only a small number of small negative eigenvalues, a useful representation of the proximity matrix may still be possible using the eigenvectors associated with the $k$ largest positive eigenvalues.

The adequacy of the resulting solution might be assessed using one of the following two criteria suggested by Mardia et al. (1979); namely

$$\frac{\sum_{i=1}^{k} |\lambda_i|}{\sum_{i=1}^{n} |\lambda_i|} \quad \text{or} \quad \frac{\sum_{i=1}^{k} \lambda_i^2}{\sum_{i=1}^{n} \lambda_i^2}.$$

Alternatively, Sibson (1979) recommends the following:

1. *Trace criterion*: Choose the number of coordinates so that the sum of their positive eigenvalues is approximately equal to the sum of all the eigenvalues.
2. *Magnitude criterion*: Accept as genuinely positive only those eigenvalues whose magnitude substantially exceeds that of the largest negative eigenvalue.

### *14.2.2 Non-metric Multidimensional Scaling*

In classical scaling the goodness-of-fit measure is based on a direct numerical comparison of observed proximities and fitted distances. In many situations however, it might be believed that the observed proximities contain little reliable information beyond that implied by their rank order. In psychological experiments, for example, proximity matrices frequently arise from asking subjects to make judgements about the similarity or dissimilarity of the stimuli of interest; in many such experiments the investigator may feel that, realistically, subjects can only give 'ordinal' judgements. For example, in comparing a range of colours they might be able to specify that one was say 'brighter' than another without being able to attach any realistic value to the extent that they differed. For such situations, what is needed is a method of multidimensional scaling, the solutions from which depend only on the rank order of the proximities, rather than their actual numerical values. In other words the solution should be invariant under monotonic transformations of the proximities. Such a method was originally suggested by Shepard (1962a,b) and Kruskal (1964a). The quintessential component of the method is the use of *monotonic regression* (see Barlow et al., 1972). In essence the aim is to represent the fitted distances, $d_{ij}$, as $d_{ij} = \hat{d}_{ij} + \varepsilon_{ij}$ where the *disparities* $\hat{d}_{ij}$ are monotonic with the observed proximities and, subject to this constraint, resemble the $d_{ij}$ as closely as possible. Algorithms to achieve this are described in Kruskal (1964b). For a given set of disparities the required coordinates can be found by minimising some function of the squared differences between the observed proximities and the derived disparities (generally known as *stress* in this context). The procedure then iterates until some convergence criterion is satisfied. Again for details see Kruskal (1964b).

## 14.3 Analysis Using R

We can apply classical scaling to the distance matrix for populations of water voles using the R function `cmdscale`. The following code finds the classical scaling solution and computes the two criteria for assessing the required number of dimensions as described above.

```
R> data("watervoles", package = "HSAUR")
R> voles_mds <- cmdscale(watervoles, k = 13, eig = TRUE)
R> voles_mds$eig
```

```
 [1]  7.359910e-01  2.626003e-01  1.492622e-01  6.990457e-02
 [5]  2.956972e-02  1.931184e-02  4.163336e-17 -1.139451e-02
 [9] -1.279569e-02 -2.849924e-02 -4.251502e-02 -5.255450e-02
[13] -7.406143e-02
```

Note that some of the eigenvalues are negative. The criterion $P_2$ can be computed by

```
R> sum(abs(voles_mds$eig[1:2]))/sum(abs(voles_mds$eig))
```

```
[1] 0.6708889
```

and the criterion suggested by Mardia et al. (1979) is

```
R> sum((voles_mds$eig[1:2])^2)/sum((voles_mds$eig)^2)
```

```
[1] 0.9391378
```

The two criteria for judging number of dimensions differ considerably, but both values are reasonably large, suggesting that the original distances between the water vole populations can be represented adequately in two dimensions. The two-dimensional solution can be plotted by extracting the coordinates from the `points` element of the `voles_mds` object; the plot is shown in Figure 14.1.

It appears that the six British populations are close to populations living in the Alps, Yugoslavia, Germany, Norway and Pyrenees I (consisting of the

```
R> x <- voles_mds$points[, 1]
R> y <- voles_mds$points[, 2]
R> plot(x, y, xlab = "Coordinate 1", ylab = "Coordinate 2",
+      xlim = range(x) * 1.2, type = "n")
R> text(x, y, labels = colnames(watervoles))
```

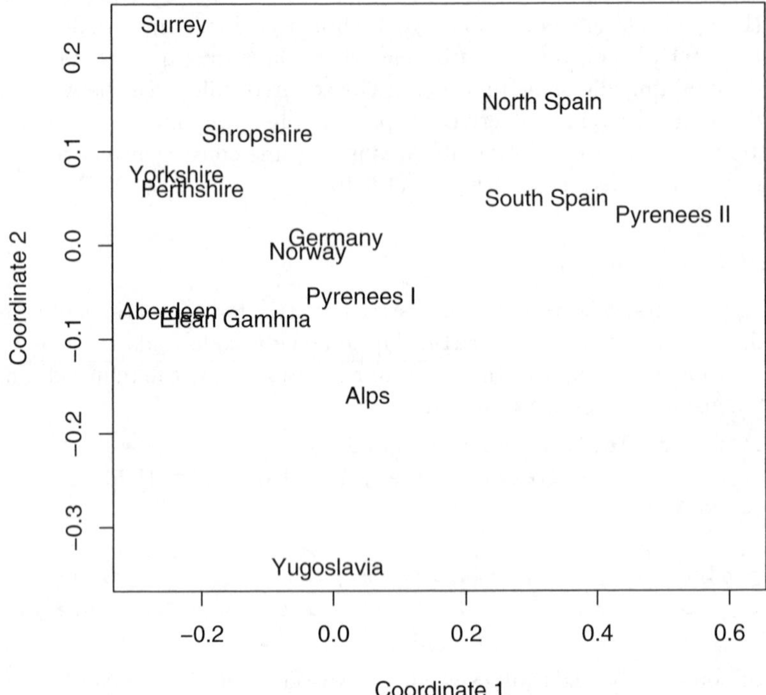


**Figure 14.1** Two-dimensional solution from classical multidimensional scaling of distance matrix for water vole populations.

species *Arvicola terrestris*) but rather distant from the populations in Pyrenees II, North Spain and South Spain (species *Arvicola sapidus*). This result would seem to imply that *Arvicola terrestris* might be present in Britain but it is less likely that this is so for *Arvicola sapidus*.

A useful graphic for highlighting possible distortions in a multidimensional scaling solution is the *minimum spanning tree*, which is defined as follows. Suppose $n$ points are given (possibly in many dimensions), then a tree spanning these points, i.e., a spanning tree, is any set of straight line segments joining pairs of points such that

- No closed loops occur,
- Every point is visited at least one time,
- The tree is connected, i.e., it has paths between any pairs of points.

The length of the tree is defined to be the sum of the length of its segments, and when a set of $n$ points and the length of all $\binom{n}{2}$ segments are given, then the minimum spanning tree is defined as the spanning tree with minimum length. Algorithms to find the minimum spanning tree of a set of $n$ points given the distances between them are given in Prim (1957) and Gower and Ross (1969).

The links of the minimum spanning tree (of the spanning tree) of the proximity matrix of interest may be plotted onto the two-dimensional scaling representation in order to identify possible distortions produced by the scaling solutions. Such distortions are indicated when nearby points on the plot are not linked by an edge of the tree.

To find the minimum spanning tree of the water vole proximity matrix, the function `mst` from package *ape* (Paradis et al., 2005) can be used and we can plot the minimum spanning tree on the two-dimensional scaling solution as shown in Figure 14.2.

The plot indicates, for example, that the apparent closeness of populations, Germany and Norway suggested by the points representing them in the MDS solution, does not reflect accurately their calculated dissimilarity; the links of the minimum spanning tree show that the Aberdeen and Elean Gamhna populations are actually more similar to the German water voles than those from Norway.

We shall now apply non-metric scaling to the voting behaviour shown in Table 14.2. Non-metric scaling is available with function `isoMDS` from package *MASS* (Venables and Ripley, 2002):

```
R> library("MASS")
R> data("voting", package = "HSAUR")
R> voting_mds <- isoMDS(voting)
```

and we again depict the two-dimensional solution (Figure 14.3). The Figure suggests that voting behaviour is essentially along party lines, although there is more variation among Republicans. The voting behaviour of one of the Republicans (Rinaldo) seems to be closer to his democratic collegues rather than to the voting behaviour of other Republicans.

The quality of a multidimensional scaling can be assessed informally by plotting the original dissimilarities and the distances obtained from a multidimensional scaling in a scatterplot, a so-called Shepard diagram. For the `voting` data, such a plot is shown in Figure 14.4. In an ideal situation, the points fall on the bisecting line, in our case, some deviations are observable.

```
R> library("ape")
R> st <- mst(watervoles)
R> plot(x, y, xlab = "Coordinate 1", ylab = "Coordinate 2",
+     xlim = range(x) * 1.2, type = "n")
R> for (i in 1:nrow(watervoles)) {
+     w1 <- which(st[i, ] == 1)
+     segments(x[i], y[i], x[w1], y[w1])
+ }
R> text(x, y, labels = colnames(watervoles))
```

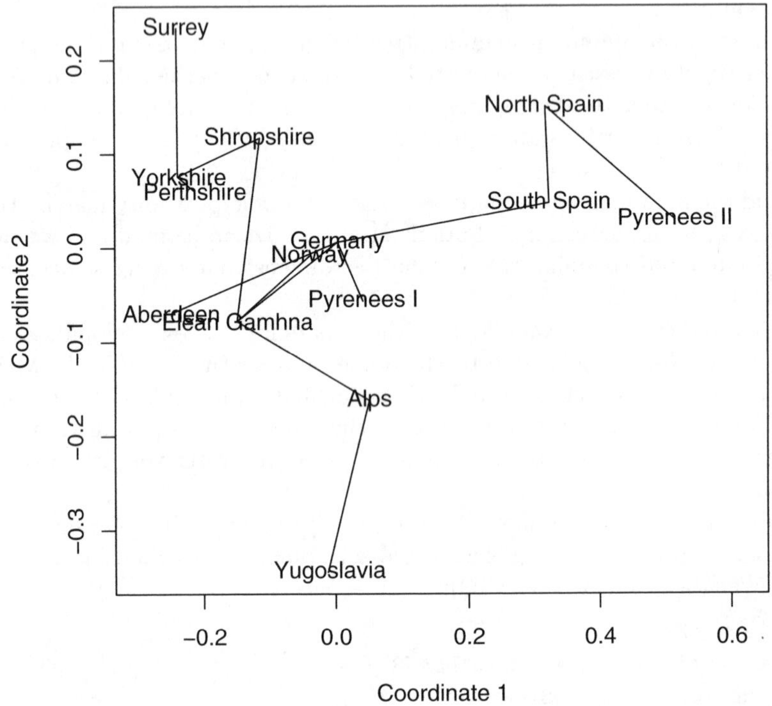


**Figure 14.2** Minimum spanning tree for the `watervoles` data.

```
R> x <- voting_mds$points[, 1]
R> y <- voting_mds$points[, 2]
R> plot(x, y, xlab = "Coordinate 1", ylab = "Coordinate 2",
+     xlim = range(voting_mds$points[, 1]) * 1.2,
+     type = "n")
R> text(x, y, labels = colnames(voting))
R> voting_sh <- Shepard(voting[lower.tri(voting)],
+     voting_mds$points)
```

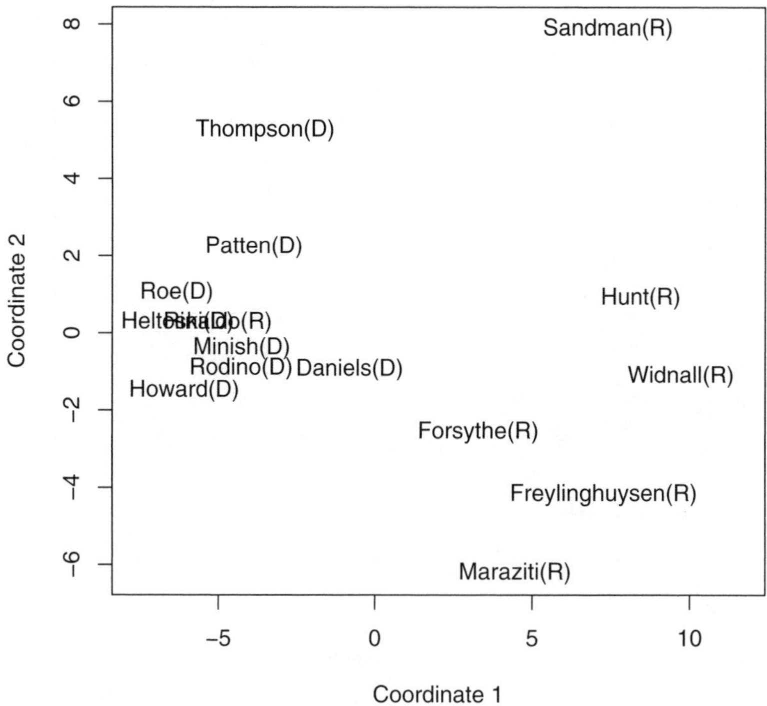


**Figure 14.3** Two-dimensional solution from non-metric multidimensional scaling of distance matrix for voting matrix.

```
R> library("MASS")
R> voting_sh <- Shepard(voting[lower.tri(voting)],
+      voting_mds$points)
R> plot(voting_sh, pch = ".", xlab = "Dissimilarity",
+      ylab = "Distance", xlim = range(voting_sh$x),
+      ylim = range(voting_sh$x))
R> lines(voting_sh$x, voting_sh$yf, type = "S")
```

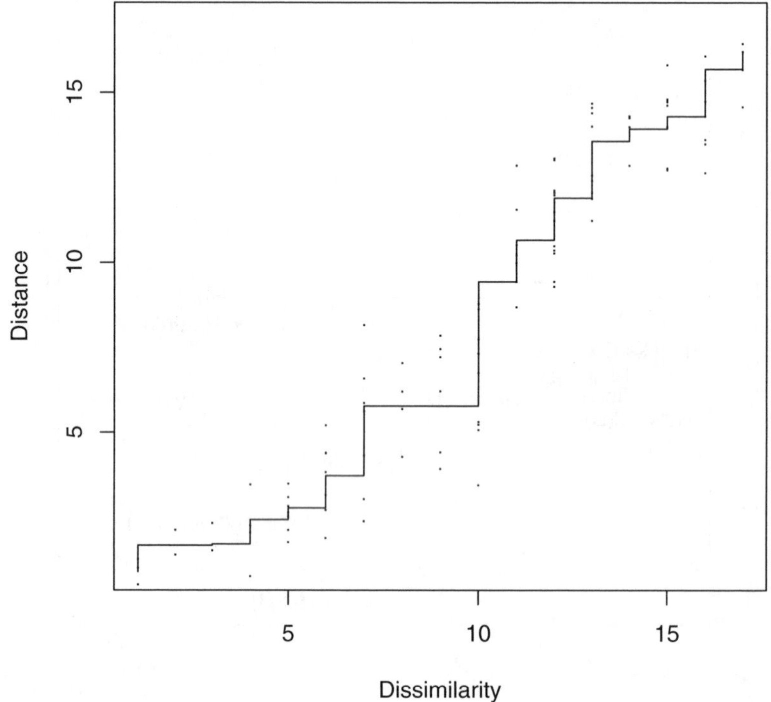


**Figure 14.4** The Shepard diagram for the `voting` data shows some discrepancies between the original dissimilarities and the multidimensional scaling solution.

## 14.4 Summary

Multidimensional scaling provides a powerful approach to extracting the structure in observed proximity matrices. Uncovering the pattern in this type of data may be important for a number of reasons, in particular for discovering the dimensions on which similarity judgements have been made.

## Exercises

Ex. 14.1 The data in Table 14.3 shows road distances between 21 European cities. Apply classical scaling to the matrix and compare the plotted two-dimensional solution with a map of Europe.

Ex. 14.2 In Table 14.4 (from Kaufman and Rousseeuw, 1990), the dissimilarity matrix of 18 species of garden flowers is shown. Use some form of multidimensional scaling to investigate which species share common properties.

Ex. 14.3 Consider 51 objects $O_1, \ldots, O_{51}$ assumed to be arranged along a straight line with the $j$th object being located at a point with coordinate $j$. Define the similarity $s_{ij}$ between object $i$ and object $j$ as

$$s_{ij} = \begin{cases} 9 & \text{if} \quad i = j \\ 8 & \text{if} \quad 1 \leq |i-j| \leq 3 \\ 7 & \text{if} \quad 4 \leq |i-j| \leq 6 \\ & \cdots \\ 1 & \text{if} \quad 22 \leq |i-j| \leq 24 \\ 0 & \text{if} \quad |i-j| \geq 25 \end{cases}$$

Convert these similarities into dissimilarities ($\delta_{ij}$) by using

$$\delta_{ij} = \sqrt{s_{ii} + s_{jj} - 2s_{ij}}$$

and then apply classical multidimensional scaling to the resulting dissimilaritiy matrix. Explain the shape of the derived two-dimensional solution.

**Table 14.3**: `eurodist` data (package *datasets*). Distances between European cities, in km.

| | Athn | Brcl | Brss | Cals | Chrb | Clgn | Cpnh | Genv | Gbrl | Hmbr | HkoH | Lsbn | Lyns | Mdrd | Mrsl | Miln | Mnch | Pars | Rome | Stck | Vinn |
|---|---|---|---|---|---|---|---|---|---|---|---|---|---|---|---|---|---|---|---|---|---|
| Athens | 0 | | | | | | | | | | | | | | | | | | | | |
| Barcelona | 3313 | 0 | | | | | | | | | | | | | | | | | | | |
| Brussels | 2963 | 1318 | 0 | | | | | | | | | | | | | | | | | | |
| Calais | 3175 | 1326 | 204 | 0 | | | | | | | | | | | | | | | | | |
| Cherbourg | 3339 | 1294 | 583 | 460 | 0 | | | | | | | | | | | | | | | | |
| Cologne | 2762 | 1498 | 206 | 409 | 785 | 0 | | | | | | | | | | | | | | | |
| Copenhagen | 3276 | 2218 | 966 | 1136 | 1545 | 760 | 0 | | | | | | | | | | | | | | |
| Geneva | 2610 | 803 | 677 | 747 | 853 | 1662 | 1418 | 0 | | | | | | | | | | | | | |
| Gibraltar | 4485 | 1172 | 2256 | 2224 | 2047 | 2436 | 3196 | 1975 | 0 | | | | | | | | | | | | |
| Hamburg | 2977 | 2018 | 597 | 714 | 1115 | 460 | 460 | 1118 | 2897 | 0 | | | | | | | | | | | |
| Hook of Holland | 3030 | 1490 | 172 | 330 | 731 | 269 | 269 | 895 | 2428 | 550 | 0 | | | | | | | | | | |
| Lisbon | 4532 | 1305 | 2084 | 2052 | 1827 | 2290 | 2971 | 1936 | 676 | 2671 | 2280 | 0 | | | | | | | | | |
| Lyons | 2753 | 645 | 690 | 739 | 789 | 714 | 1458 | 158 | 1817 | 1159 | 863 | 1178 | 0 | | | | | | | | |
| Madrid | 3949 | 636 | 1558 | 1550 | 1347 | 1764 | 2498 | 1439 | 698 | 2198 | 1730 | 668 | 1281 | 0 | | | | | | | |
| Marseilles | 2865 | 521 | 1011 | 1059 | 1101 | 1035 | 1778 | 425 | 1693 | 1479 | 1183 | 1762 | 320 | 1157 | 0 | | | | | | |
| Milan | 2282 | 1014 | 925 | 1077 | 1209 | 911 | 1537 | 328 | 2185 | 1238 | 1098 | 2250 | 328 | 1724 | 618 | 0 | | | | | |
| Munich | 2179 | 1365 | 747 | 977 | 1160 | 583 | 1104 | 591 | 2565 | 805 | 851 | 2507 | 724 | 2010 | 1109 | 331 | 0 | | | | |
| Paris | 3000 | 1033 | 285 | 280 | 340 | 465 | 1176 | 513 | 1971 | 877 | 457 | 1799 | 471 | 1273 | 792 | 856 | 821 | 0 | | | |
| Rome | 817 | 1460 | 1511 | 1662 | 1794 | 1497 | 2050 | 995 | 2631 | 1751 | 1683 | 2700 | 1048 | 2097 | 1011 | 586 | 946 | 1476 | 0 | | |
| Stockholm | 3927 | 2868 | 1616 | 1786 | 2196 | 1403 | 650 | 2068 | 3886 | 949 | 1500 | 3231 | 2108 | 3188 | 2428 | 2187 | 1754 | 1827 | 2707 | 0 | |
| Vienna | 1991 | 1802 | 1175 | 1381 | 1588 | 937 | 1455 | 1019 | 2974 | 1155 | 1205 | 2937 | 1157 | 2409 | 1363 | 898 | 428 | 1249 | 1209 | 2105 | 0 |

**Table 14.4**: `gardenflowers` data. Dissimilarity matrix of 18 species of gardenflowers.

| | Bgn | Brm | Cml | Dhl | F- | Fch | Grn | Gld | Hth | Hyd | Irs | Lly | L- | Pny | Pnc | Rdr | Scr | Tlp |
|---|---|---|---|---|---|---|---|---|---|---|---|---|---|---|---|---|---|---|
| Begonia | 0.00 | | | | | | | | | | | | | | | | | |
| Broom | 0.91 | 0.00 | | | | | | | | | | | | | | | | |
| Camellia | 0.49 | 0.67 | 0.00 | | | | | | | | | | | | | | | |
| Dahlia | 0.47 | 0.59 | 0.59 | 0.00 | | | | | | | | | | | | | | |
| Forget-me-not | 0.43 | 0.90 | 0.57 | 0.61 | 0.00 | | | | | | | | | | | | | |
| Fuchsia | 0.23 | 0.79 | 0.29 | 0.52 | 0.44 | 0.00 | | | | | | | | | | | | |
| Geranium | 0.31 | 0.70 | 0.54 | 0.44 | 0.54 | 0.24 | 0.00 | | | | | | | | | | | |
| Gladiolus | 0.49 | 0.57 | 0.71 | 0.26 | 0.49 | 0.68 | 0.49 | 0.00 | | | | | | | | | | |
| Heather | 0.57 | 0.57 | 0.57 | 0.89 | 0.50 | 0.61 | 0.70 | 0.77 | 0.00 | | | | | | | | | |
| Hydrangae | 0.76 | 0.58 | 0.58 | 0.62 | 0.39 | 0.61 | 0.86 | 0.70 | 0.55 | 0.00 | | | | | | | | |
| Iris | 0.32 | 0.77 | 0.63 | 0.75 | 0.46 | 0.52 | 0.60 | 0.63 | 0.46 | 0.47 | 0.00 | | | | | | | |
| Lily | 0.51 | 0.69 | 0.69 | 0.53 | 0.51 | 0.65 | 0.77 | 0.47 | 0.51 | 0.39 | 0.36 | 0.00 | | | | | | |
| Lily-of-the-valley | 0.59 | 0.75 | 0.75 | 0.77 | 0.35 | 0.63 | 0.72 | 0.65 | 0.35 | 0.41 | 0.45 | 0.24 | 0.00 | | | | | |
| Peony | 0.37 | 0.68 | 0.68 | 0.38 | 0.52 | 0.48 | 0.63 | 0.49 | 0.52 | 0.39 | 0.37 | 0.17 | 0.39 | 0.00 | | | | |
| Pink carnation | 0.74 | 0.54 | 0.70 | 0.58 | 0.54 | 0.74 | 0.50 | 0.49 | 0.36 | 0.52 | 0.60 | 0.48 | 0.39 | 0.49 | 0.00 | | | |
| Red rose | 0.84 | 0.41 | 0.75 | 0.37 | 0.82 | 0.71 | 0.61 | 0.64 | 0.81 | 0.43 | 0.84 | 0.62 | 0.67 | 0.47 | 0.45 | 0.00 | | |
| Scotch rose | 0.94 | 0.20 | 0.70 | 0.48 | 0.77 | 0.83 | 0.74 | 0.45 | 0.77 | 0.38 | 0.80 | 0.58 | 0.62 | 0.57 | 0.40 | 0.21 | 0.00 | |
| Tulip | 0.44 | 0.50 | 0.79 | 0.48 | 0.59 | 0.68 | 0.47 | 0.22 | 0.59 | 0.92 | 0.59 | 0.67 | 0.72 | 0.67 | 0.61 | 0.85 | 0.67 | 0.00 |

CHAPTER 15

# Cluster Analysis: Classifying the Exoplanets

## 15.1 Introduction

Exoplanets are planets outside the Solar System. The first such planet was discovered in 1995 by Mayor and Queloz (1995). The planet, similar in mass to Jupiter, was found orbiting a relatively ordinary star, 51 Pegasus. In the intervening period over a hundred exoplanets have been discovered, nearly all being detected indirectly, using the gravitational influence they exert on their associated central stars. A fascinating account of exoplanets and their discovery is given in Mayor and Frei (2003).

From the properties of the exoplanets found up to now it appears that the theory of planetary development constructed for the planets of the Solar System may need to be reformulated. The exoplanets are not at all like the nine local planets that we know so well. A first step in the process of understanding the exoplanets might be to try to classify them with respect to their known properties and this will be the aim in this chapter. The data in Table 15.1 (taken with permission from Mayor and Frei, 2003), gives the mass (in Jupiter mass, `mass`), the period (in earth days, `period`) and the eccentricity (`eccent`) of the exoplanets discovered up until October 2002. We shall investigate the structure of these data using several methods of *cluster analysis*.

**Table 15.1**: `planets` data. Jupiter mass, period and eccentricity of exoplanets.

| mass | period | eccen | mass | period | eccen |
|---|---|---|---|---|---|
| 0.120 | 4.950000 | 0.0000 | 1.890 | 61.020000 | 0.1000 |
| 0.197 | 3.971000 | 0.0000 | 1.900 | 6.276000 | 0.1500 |
| 0.210 | 44.280000 | 0.3400 | 1.990 | 743.000000 | 0.6200 |
| 0.220 | 75.800000 | 0.2800 | 2.050 | 241.300000 | 0.2400 |
| 0.230 | 6.403000 | 0.0800 | 0.050 | 1119.000000 | 0.1700 |
| 0.250 | 3.024000 | 0.0200 | 2.080 | 228.520000 | 0.3040 |
| 0.340 | 2.985000 | 0.0800 | 2.240 | 311.300000 | 0.2200 |
| 0.400 | 10.901000 | 0.4980 | 2.540 | 1089.000000 | 0.0600 |
| 0.420 | 3.509700 | 0.0000 | 2.540 | 627.340000 | 0.0600 |
| 0.470 | 4.229000 | 0.0000 | 2.550 | 2185.000000 | 0.1800 |
| 0.480 | 3.487000 | 0.0500 | 2.630 | 414.000000 | 0.2100 |
| 0.480 | 22.090000 | 0.3000 | 2.840 | 250.500000 | 0.1900 |

**Table 15.1**: `planets` data (continued).

| mass | period | eccen | mass | period | eccen |
|---|---|---|---|---|---|
| 0.540 | 3.097000 | 0.0100 | 2.940 | 229.900000 | 0.3500 |
| 0.560 | 30.120000 | 0.2700 | 3.030 | 186.900000 | 0.4100 |
| 0.680 | 4.617000 | 0.0200 | 3.320 | 267.200000 | 0.2300 |
| 0.685 | 3.524330 | 0.0000 | 3.360 | 1098.000000 | 0.2200 |
| 0.760 | 2594.000000 | 0.1000 | 3.370 | 133.710000 | 0.5110 |
| 0.770 | 14.310000 | 0.2700 | 3.440 | 1112.000000 | 0.5200 |
| 0.810 | 828.950000 | 0.0400 | 3.550 | 18.200000 | 0.0100 |
| 0.880 | 221.600000 | 0.5400 | 3.810 | 340.000000 | 0.3600 |
| 0.880 | 2518.000000 | 0.6000 | 3.900 | 111.810000 | 0.9270 |
| 0.890 | 64.620000 | 0.1300 | 4.000 | 15.780000 | 0.0460 |
| 0.900 | 1136.000000 | 0.3300 | 4.000 | 5360.000000 | 0.1600 |
| 0.930 | 3.092000 | 0.0000 | 4.120 | 1209.900000 | 0.6500 |
| 0.930 | 14.660000 | 0.0300 | 4.140 | 3.313000 | 0.0200 |
| 0.990 | 39.810000 | 0.0700 | 4.270 | 1764.000000 | 0.3530 |
| 0.990 | 500.730000 | 0.1000 | 4.290 | 1308.500000 | 0.3100 |
| 0.990 | 872.300000 | 0.2800 | 4.500 | 951.000000 | 0.4500 |
| 1.000 | 337.110000 | 0.3800 | 4.800 | 1237.000000 | 0.5150 |
| 1.000 | 264.900000 | 0.3800 | 5.180 | 576.000000 | 0.7100 |
| 1.010 | 540.400000 | 0.5200 | 5.700 | 383.000000 | 0.0700 |
| 1.010 | 1942.000000 | 0.4000 | 6.080 | 1074.000000 | 0.0110 |
| 1.020 | 10.720000 | 0.0440 | 6.292 | 71.487000 | 0.1243 |
| 1.050 | 119.600000 | 0.3500 | 7.170 | 256.000000 | 0.7000 |
| 1.120 | 500.000000 | 0.2300 | 7.390 | 1582.000000 | 0.4780 |
| 1.130 | 154.800000 | 0.3100 | 7.420 | 116.700000 | 0.4000 |
| 1.150 | 2614.000000 | 0.0000 | 7.500 | 2300.000000 | 0.3950 |
| 1.230 | 1326.000000 | 0.1400 | 7.700 | 58.116000 | 0.5290 |
| 1.240 | 391.000000 | 0.4000 | 7.950 | 1620.000000 | 0.2200 |
| 1.240 | 435.600000 | 0.4500 | 8.000 | 1558.000000 | 0.3140 |
| 1.282 | 7.126200 | 0.1340 | 8.640 | 550.650000 | 0.7100 |
| 1.420 | 426.000000 | 0.0200 | 9.700 | 653.220000 | 0.4100 |
| 1.550 | 51.610000 | 0.6490 | 10.000 | 3030.000000 | 0.5600 |
| 1.560 | 1444.500000 | 0.2000 | 10.370 | 2115.200000 | 0.6200 |
| 1.580 | 260.000000 | 0.2400 | 10.960 | 84.030000 | 0.3300 |
| 1.630 | 444.600000 | 0.4100 | 11.300 | 2189.000000 | 0.3400 |
| 1.640 | 406.000000 | 0.5300 | 11.980 | 1209.000000 | 0.3700 |
| 1.650 | 401.100000 | 0.3600 | 14.400 | 8.428198 | 0.2770 |
| 1.680 | 796.700000 | 0.6800 | 16.900 | 1739.500000 | 0.2280 |
| 1.760 | 903.000000 | 0.2000 | 17.500 | 256.030000 | 0.4290 |
| 1.830 | 454.000000 | 0.2000 | | | |

*Source*: From Mayor, M., Frei, P.-Y., and Roukema, B., *New Worlds in the Cosmos*, Cambridge University Press, Cambridge, England, 2003. With permission.

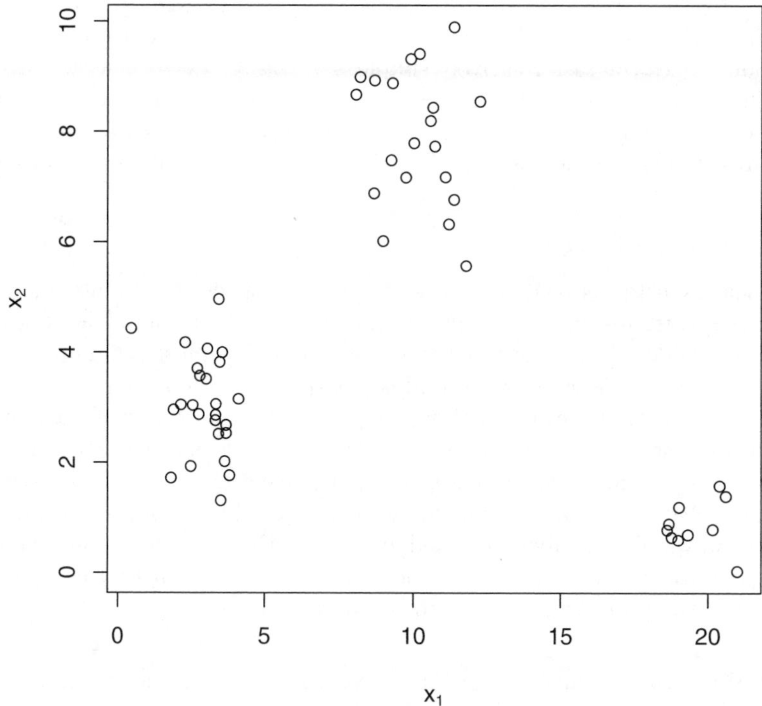


**Figure 15.1** Bivariate data showing the presence of three clusters.

## 15.2 Cluster Analysis

Cluster analysis is a generic term for a wide range of numerical methods for examining multivariate data with a view to uncovering or discovering groups or clusters of observations that are homogeneous and separated from other groups. In medicine, for example, discovering that a sample of patients with measurements on a variety of characteristics and symptoms, actually consists of a small number of groups within which these characteristics are relatively similar, and between which they are different, might have important implications both in terms of future treatment and for investigating the aetiology of a condition. More recently cluster analysis techniques have been applied to microarray data (Alon et al., 1999, among many others), image analysis (Everitt and Bullmore, 1999) or in marketing science (Dolnicar and Leisch, 2003).

Clustering techniques essentially try to formalise what human observers do so well in two or three dimensions. Consider, for example, the scatterplot shown in Figure 15.1. The conclusion that there are three natural groups or clusters of dots is reached with no conscious effort or thought. Clusters are identified by the assessment of the relative distances between points and in this example, the relative homogeneity of each cluster and the degree of their separation makes the task relatively simple.

Detailed accounts of clustering techniques are available in Everitt et al. (2001) and Gordon (1999). Here we concentrate on two types of clustering procedures: $k$-means type and classification maximum likelihood methods.

### *15.2.1 k-Means Clustering*

The $k$-means clustering technique seeks to partition a set of data into a specified number of groups, $k$, by minimising some numerical criterion, low values of which are considered indicative of a 'good' solution. The most commonly used approach, for example, is to try to find the partition of the $n$ individuals into $k$ groups, which minimises the within-group sum of squares over all variables. The problem then appears relatively simple; namely, consider every possible partition of the $n$ individuals into $k$ groups, and select the one with the lowest within-group sum of squares. Unfortunately, the problem in practice is not so straightforward. The numbers involved are so vast that complete enumeration of *every* possible partition remains impossible even with the fastest computer. The scale of the problem is illustrated by the numbers in Table 15.2.

**Table 15.2**: Number of possible partitions depending on the sample size $n$ and number of clusters $k$.

| $n$ | $k$ | Number of possible partitions |
|---|---|---|
| 15 | 3 | 2,375,101 |
| 20 | 4 | 45,232,115,901 |
| 25 | 8 | 690,223,721,118,368,580 |
| 100 | 5 | $10^{68}$ |

The impracticability of examining every possible partition has led to the development of algorithms designed to search for the minimum values of the clustering criterion by rearranging existing partitions and keeping the new one only if it provides an improvement. Such algorithms do not, of course, guarantee finding the global minimum of the criterion. The essential steps in these algorithms are as follows:

1. Find some initial partition of the individuals into the required number of groups. Such an initial partition could be provided by a solution from one of the hierarchical clustering techniques described in the previous section.
2. Calculate the change in the clustering criterion produced by 'moving' each individual from its own to another cluster.

3. Make the change that leads to the greatest improvement in the value of the clustering criterion.
4. Repeat steps 2 and 3 until no move of an individual causes the clustering criterion to improve.

When variables are on very different scales (as they are for the exoplanets data) some form of standardization will be needed before applying $k$-means clustering (for a detailed discussion of this problem see Everitt et al., 2001).

### *15.2.2 Model-based Clustering*

The $k$-means clustering method described in the previous section is based largely in heuristic but intuitively reasonable procedures. But it is not based on formal models thus making problems such as deciding on a particular method, estimating the number of clusters, etc., particularly difficult. And, of course, without a reasonable model, formal inference is precluded. In practice these may not be insurmountable objections to the use of the technique since cluster analysis is essentially an 'exploratory' tool. But model-based cluster methods do have some advantages, and a variety of possibilities have been proposed. The most successful approach has been that proposed by Scott and Symons (1971) and extended by Banfield and Raftery (1993) and Fraley and Raftery (1999, 2002), in which it is assumed that the population from which the observations arise consists of $c$ subpopulations each corresponding to a cluster, and that the density of a $q$-dimensional observation $\mathbf{x}^\top = (x_1, \dots, x_q)$ from the $j$th subpopulation is $f_j(\mathbf{x}, \vartheta_j), j = 1, \dots, c$, for some unknown vector of parameters, $\vartheta_j$. They also introduce a vector $\gamma = (\gamma_1, \dots, \gamma_n)$, where $\gamma_i = j$ of $\mathbf{x}_i$ is from the $j$ subpopulation. The $\gamma_i$ label the subpopulation for each observation $i = 1, \dots, n$. The clustering problem now becomes that of choosing $\vartheta = (\vartheta_1, \dots, \vartheta_c)$ and $\gamma$ to maximise the likelihood function associated with such assumptions. This classification maximum likelihood procedure is described briefly in the sequel.

### *15.2.3 Classification Maximum Likelihood*

Assume the population consists of $c$ subpopulations, each corresponding to a cluster of observations, and that the density function of a $q$-dimensional observation from the $j$th subpopulation is $f_j(\mathbf{x}, \vartheta_j)$ for some unknown vector of parameters, $\vartheta_j$. Also, assume that $\gamma = (\gamma_1, \dots, \gamma_n)$ gives the labels of the subpopulation to which the observation belongs: so $\gamma_i = j$ if $\mathbf{x}_i$ is from the $j$th population.

The clustering problem becomes that of choosing $\vartheta = (\vartheta_1, \dots, \vartheta_c)$ and $\gamma$ to maximise the likelihood

$$L(\vartheta, \gamma) = \prod_{i=1}^{n} f_{\gamma_i}(\mathbf{x}_i, \vartheta_{\gamma_i}). \tag{15.1}$$

If $f_j(\mathbf{x}, \vartheta_j)$ is taken as the multivariate normal density with mean vector $\mu_j$

and covariance matrix $\Sigma_j$, this likelihood has the form

$$L(\vartheta, \gamma) = \prod_{j=1}^{c} \prod_{i:\gamma_i=j} |\Sigma_j|^{1/2} \exp\left(-\frac{1}{2}(\mathbf{x}_i - \mu_j)^\top \Sigma_j^{-1}(\mathbf{x}_i - \mu_j)\right). \qquad (15.2)$$

The maximum likelihood estimator of $\mu_j$ is $\hat{\mu}_j = n_j^{-1} \sum_{i:\gamma_i=j} \mathbf{x}_i$ where the number of observations in each subpopulation is $n_j = \sum_{i=1}^{n} I(\gamma_i = j)$. Replacing $\mu_j$ in (15.2) yields the following log-likelihood

$$l(\vartheta, \gamma) = -\frac{1}{2} \sum_{j=1}^{c} \text{trace}(\mathbf{W}_j \Sigma_j^{-1} + n \log |\Sigma_j|)$$

where $\mathbf{W}_j$ is the $p \times p$ matrix of sums of squares and cross-products of the variables for subpopulation $j$.

Banfield and Raftery (1993) demonstrate the following: If the covariance matrix $\Sigma_j$ is $\sigma^2$ times the identity matrix for all populations $j = 1, \dots, c$, then the likelihood is maximised by choosing $\gamma$ to minimise trace($\mathbf{W}$), where $\mathbf{W} = \sum_{j=1}^{c} \mathbf{W}_j$, i.e., minimisation of the written group sum of squares. Use of this criterion in a cluster analysis will lend to produce spherical clusters of largely equal sizes.

If $\Sigma_k = \Sigma$ for $j = 1, \dots, c$, then the likelihood is maximised by choosing $\gamma$ to minimise $|\mathbf{W}|$, a clustering criterion discussed by Friedman and Rubin (1967) and Marriott (1982). Use of this criterion in a cluster analysis will lend to produce clusters with the same elliptical slope.

If $\Sigma_j$ is not constrained, the likelihood is maximised by choosing $\gamma$ to minimise $\sum_{j=1}^{c} n_j \log |\mathbf{W}_j|/n_j$.

Banfield and Raftery (1993) also consider criteria that allow the shape of clusters to less constrained than with the minimisation of trace($\mathbf{W}$) and $|\mathbf{W}|$ criteria, but that remain more parsimonious than the completely unconstrained model. For example, constraining clusters to be spherical but not to have the same volume, or constraining clusters to have diagonal covariance matrices but allowing their shapes, sizes and orientations to vary.

The EM algorithm (see Dempster et al., 1977), is used for maximum likelihood estimation – details are given in Fraley and Raftery (1999). Model selection is a combination of choosing the appropriate clustering model and the optimal number of clusters. A Bayesian approach is used (see Fraley and Raftery, 1999), using what is known as the *Bayesian Information Criterion* (BIC).

## 15.3 Analysis Using R

Prior to a cluster analysis we present a graphical representation of the three-dimensional `planets` data by means of the *scatterplot3d* package (Ligges and Mächler, 2003). The logarithms of the mass, period and eccentricity measurements are shown in a scatterplot in Figure 15.2. The diagram gives no clear indication of distinct clusters in the data but nevertheless we shall continue to investigate this possibility by applying $k$-means clustering with the `kmeans` function in R. In essence this method finds a partition of the observations for a particular number of clusters by minimizing the total within-group sum of squares over all variables. Deciding on the 'optimal' number of groups is often difficult and there is no method that can be recommended in all circumstances (see Everitt et al., 2001). An informal approach to the number of groups problem is to plot the within-group sum of squares for each partition given by applying the `kmeans` procedure and looking for an 'elbow' in the resulting curve (cf. scree plots in factor analysis). Such a plot can be constructed in R for the `planets` data using the code displayed with Figure 15.3 (note that since the three variables are on very different scales they first need to be standardised in some way – here we use the range of each).

Sadly Figure 15.3 gives no completely convincing verdict on the number of groups we should consider, but using a little imagination 'little elbows' can be spotted at the three and five group solutions. We can find the number of planets in each group using

```
R> planet_kmeans3 <- kmeans(planet.dat, centers = 3)
R> table(planet_kmeans3$cluster)

 1  2  3
34 14 53
```

The centers of the clusters for the untransformed data can be computed using a small convenience function

```
R> ccent <- function(cl) {
+     f <- function(i) colMeans(planets[cl == i, ])
+     x <- sapply(sort(unique(cl)), f)
+     colnames(x) <- sort(unique(cl))
+     return(x)
+ }
```

which, applied to the three cluster solution obtained by $k$-means gets

```
R> ccent(planet_kmeans3$cluster)

                 1          2           3
mass       2.9276471   10.56786   1.6710566
period   616.0760882 1693.17201 427.7105892
eccen      0.4953529    0.36650   0.1219491
```

for the three cluster solution and, for the five cluster solution using

```
R> planet_kmeans5 <- kmeans(planet.dat, centers = 5)
R> table(planet_kmeans5$cluster)

 1  2  3  4  5
36 16  4 29 16

R> ccent(planet_kmeans5$cluster)

                1           2        3           4
mass      0.9103889   3.4895000    2.115   2.4579310
```

```
R> data("planets", package = "HSAUR")
R> library("scatterplot3d")
R> scatterplot3d(log(planets$mass), log(planets$period),
+     log(planets$eccen), type = "h", angle = 55,
+     scale.y = 0.7, pch = 16, y.ticklabs = seq(0,
+         10, by = 2), y.margin.add = 0.1)
```

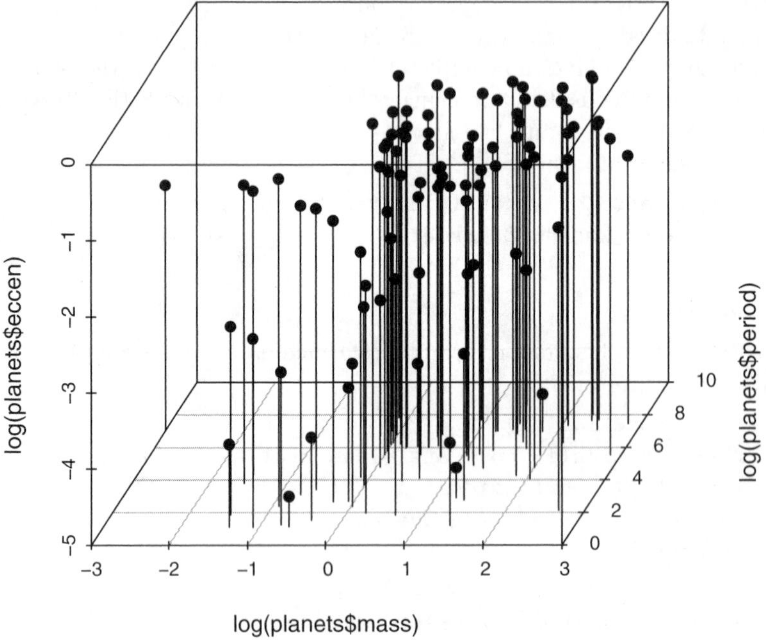


**Figure 15.2** 3D scatterplot of the logarithms of the three variables available for each of the exoplanets.

```
period 175.9617008 643.3637500 3188.250 650.6324483
eccen    0.1314444   0.1313313    0.110   0.5018276
                 5
mass     10.481875
period 1191.867137
eccen     0.413125
```

Interpretation of both the three and five cluster solutions clearly requires a detailed knowledge of astronomy. But the mean vectors of the three group

```
R> rge <- apply(planets, 2, max) - apply(planets, 2,
+     min)
R> planet.dat <- sweep(planets, 2, rge, FUN = "/")
R> n <- nrow(planet.dat)
R> wss <- rep(0, 10)
R> wss[1] <- (n - 1) * sum(apply(planet.dat, 2, var))
R> for (i in 2:10) wss[i] <- sum(kmeans(planet.dat,
+     centers = i)$withinss)
R> plot(1:10, wss, type = "b", xlab = "Number of groups",
+     ylab = "Within groups sum of squares")
```

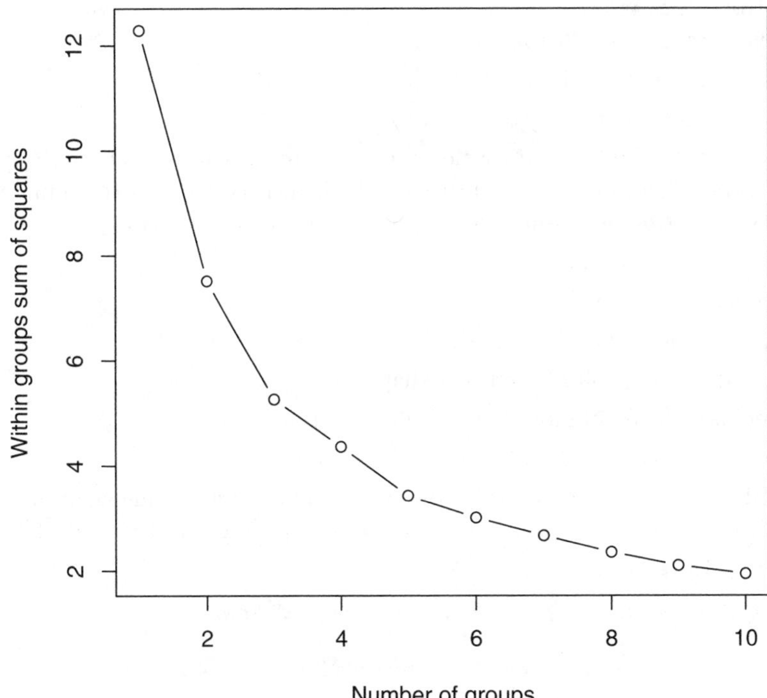


**Figure 15.3** Within-cluster sum of squares for different numbers of clusters for the exoplanet data.

solution, for example, imply a relatively large class of Jupiter-sized planets with small periods and small eccentricities, a smaller class of massive planets with moderate periods and large eccentricities, and a very small class of large planets with extreme periods and moderate eccentricities.

### *15.3.1 Model-based Clustering in R*

We now proceed to apply model-based clustering to the planets data. R functions for model-based clustering are available in package *mclust* (Fraley et al., 2005, Fraley and Raftery, 2002). Here we use the `Mclust` function since this selects both the most appropriate model for the data *and* the optimal number of groups based on the values of the BIC computed over several models and a range of values for number of groups. The necessary code is:

```
R> library("mclust")
R> planet_mclust <- Mclust(planet.dat)
```

and we first examine a plot of BIC values using

```
R> plot(planet_mclust, planet.dat)
```

and selecting the BIC option (option number 1 to be selected interactively). The resulting diagram is shown in Figure 15.4. In this diagram the numbers refer to different model assumptions about the shape of clusters:

1. Spherical, equal volume,
2. Spherical, unequal volume,
3. Diagonal equal volume, equal shape,
4. Diagonal varying volume, varying shape,
5. Ellipsoidal, equal volume, shape and orientation,
6. Ellipsoidal, varying volume, shape and orientation.

The BIC selects model 4 (diagonal varying volume and varying shape) with three clusters as the best solution as can be seen from the `print` output:

```
R> print(planet_mclust)

 best model: diagonal, varying volume and shape with 3 groups

 averge/median classification uncertainty: 0.043 / 0.012
```

This solution can be shown graphically as a scatterplot matrix The plot is shown in Figure 15.5. Figure 15.6 depicts the clustering solution in the three-dimensional space.

The number of planets in each cluster and the mean vectors of the three clusters for the untransformed data can now be inspected by using

```
R> table(planet_mclust$classification)

 1  2  3
19 41 41

R> ccent(planet_mclust$classification)
```

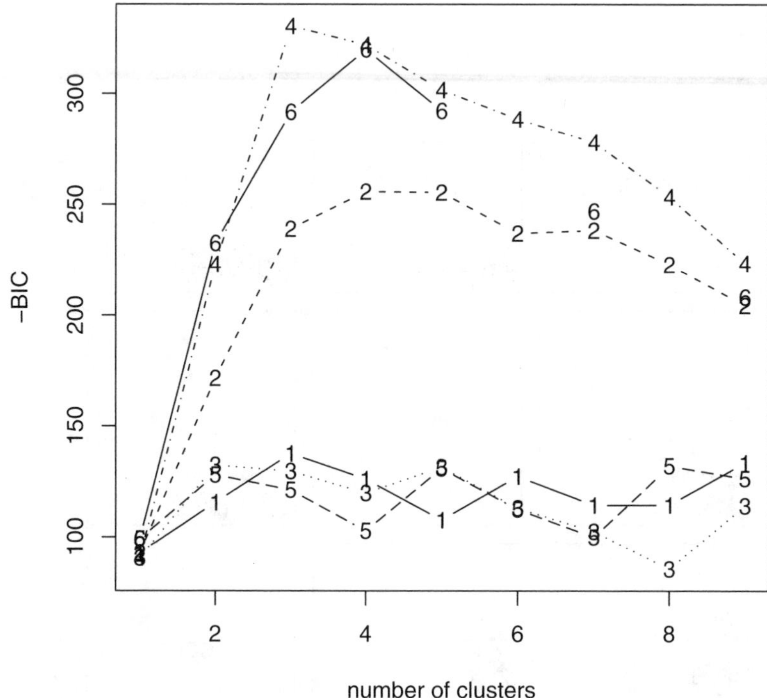


**Figure 15.4** Plot of BIC values for a variety of models and a range of number of clusters.

```
                1           2            3
mass   1.16652632   1.5797561    6.0761463
period 6.47180158 313.4127073 1325.5310048
eccen  0.03652632   0.3061463    0.3704951
```

Cluster 1 consists of planets about the same size as Jupiter with very short periods and eccentricities (similar to the first cluster of the $k$-means solution). Cluster 2 consists of slightly larger planets with moderate periods and large eccentricities, and cluster 3 contains the very large planets with very large periods. These two clusters do not match those found by the $k$-means approach.

## 15.4 Summary

Cluster analysis techniques provide a rich source of possible strategies for exploring complex multivariate data. But the use of cluster analysis in practice

```
R> x <- clPairs(planet.dat,
+      classification = planet_mclust$classification,
+      symbols = 1:3, col = "black")
```

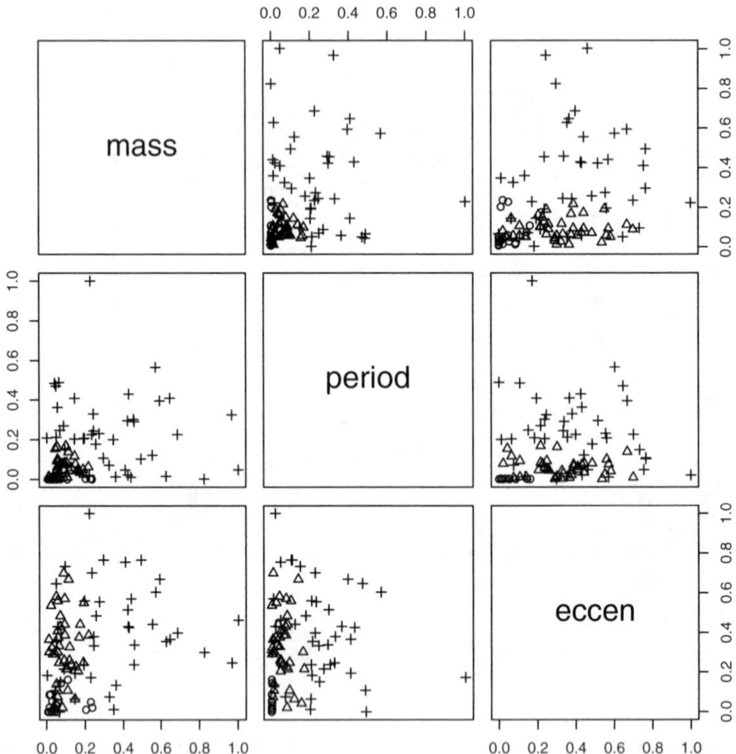


**Figure 15.5** Scatterplot matrix of planets data showing a three cluster solution from `Mclust`.

does not involve simply the application of one particular technique to the data under investigation, but rather necessitates a series of steps, each of which may be dependent on the results of the preceding one. It is generally impossible a priori to anticipate what combination of variables, similarity measures and clustering technique is likely to lead to interesting and informative classifications. Consequently, the analysis proceeds through several stages, with the researcher intervening if necessary to alter variables, choose a different similarity measure, concentrate on a particular subset of individuals, and so on. The final, extremely important, stage concerns the evaluation of the clustering solutions obtained. Are the clusters 'real' or merely artefacts of the algorithms?

```
R> scatterplot3d(log(planets$mass), log(planets$period),
+      log(planets$eccen), type = "h", angle = 55,
+      scale.y = 0.7, pch = planet_mclust$classification,
+      y.ticklabs = seq(0, 10, by = 2), y.margin.add = 0.1)
```

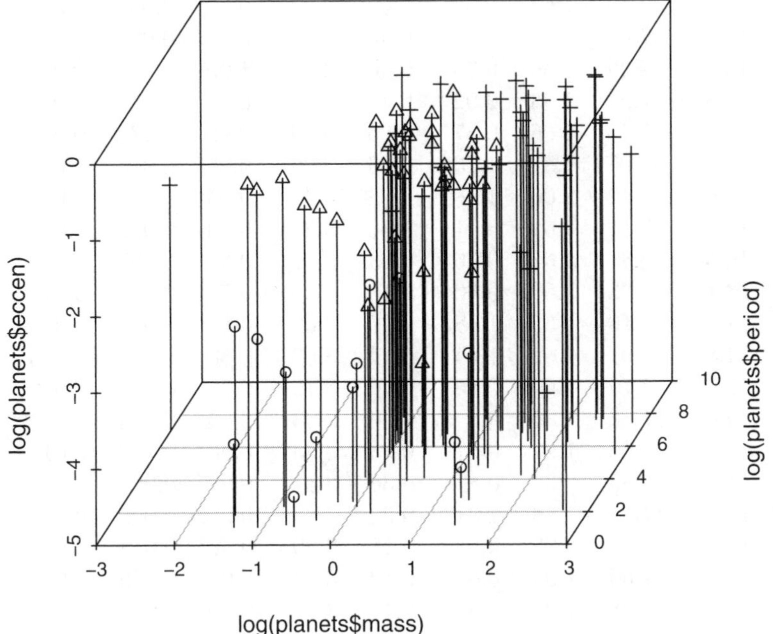


**Figure 15.6** 3D scatterplot of planets data showing a three cluster solution from `Mclust`.

Do other solutions exist that are better in some sense? Can the clusters be given a convincing interpretation? A long list of such questions might be posed, and readers intending to apply clustering to their data are recommended to read the detailed accounts of cluster evaluation given in Dubes and Jain (1979) and in Everitt et al. (2001).

## Exercises

Ex. 15.1 The data shown in Table 15.3 give the chemical composition of 48 specimens of Romano-British pottery, determined by atomic absorption spectrophotometry, for nine oxides (Tubb et al., 1980). Analyse the pottery data using `Mclust`. To what model in `Mclust` does the $k$-mean approach approximate?

**Table 15.3**: `pottery` data. Romano-British pottery data.

| Al2O3 | Fe2O3 | MgO | CaO | Na2O | K2O | TiO2 | MnO | BaO |
|---|---|---|---|---|---|---|---|---|
| 18.8 | 9.52 | 2.00 | 0.79 | 0.40 | 3.20 | 1.01 | 0.077 | 0.015 |
| 16.9 | 7.33 | 1.65 | 0.84 | 0.40 | 3.05 | 0.99 | 0.067 | 0.018 |
| 18.2 | 7.64 | 1.82 | 0.77 | 0.40 | 3.07 | 0.98 | 0.087 | 0.014 |
| 16.9 | 7.29 | 1.56 | 0.76 | 0.40 | 3.05 | 1.00 | 0.063 | 0.019 |
| 17.8 | 7.24 | 1.83 | 0.92 | 0.43 | 3.12 | 0.93 | 0.061 | 0.019 |
| 18.8 | 7.45 | 2.06 | 0.87 | 0.25 | 3.26 | 0.98 | 0.072 | 0.017 |
| 16.5 | 7.05 | 1.81 | 1.73 | 0.33 | 3.20 | 0.95 | 0.066 | 0.019 |
| 18.0 | 7.42 | 2.06 | 1.00 | 0.28 | 3.37 | 0.96 | 0.072 | 0.017 |
| 15.8 | 7.15 | 1.62 | 0.71 | 0.38 | 3.25 | 0.93 | 0.062 | 0.017 |
| 14.6 | 6.87 | 1.67 | 0.76 | 0.33 | 3.06 | 0.91 | 0.055 | 0.012 |
| 13.7 | 5.83 | 1.50 | 0.66 | 0.13 | 2.25 | 0.75 | 0.034 | 0.012 |
| 14.6 | 6.76 | 1.63 | 1.48 | 0.20 | 3.02 | 0.87 | 0.055 | 0.016 |
| 14.8 | 7.07 | 1.62 | 1.44 | 0.24 | 3.03 | 0.86 | 0.080 | 0.016 |
| 17.1 | 7.79 | 1.99 | 0.83 | 0.46 | 3.13 | 0.93 | 0.090 | 0.020 |
| 16.8 | 7.86 | 1.86 | 0.84 | 0.46 | 2.93 | 0.94 | 0.094 | 0.020 |
| 15.8 | 7.65 | 1.94 | 0.81 | 0.83 | 3.33 | 0.96 | 0.112 | 0.019 |
| 18.6 | 7.85 | 2.33 | 0.87 | 0.38 | 3.17 | 0.98 | 0.081 | 0.018 |
| 16.9 | 7.87 | 1.83 | 1.31 | 0.53 | 3.09 | 0.95 | 0.092 | 0.023 |
| 18.9 | 7.58 | 2.05 | 0.83 | 0.13 | 3.29 | 0.98 | 0.072 | 0.015 |
| 18.0 | 7.50 | 1.94 | 0.69 | 0.12 | 3.14 | 0.93 | 0.035 | 0.017 |
| 17.8 | 7.28 | 1.92 | 0.81 | 0.18 | 3.15 | 0.90 | 0.067 | 0.017 |
| 14.4 | 7.00 | 4.30 | 0.15 | 0.51 | 4.25 | 0.79 | 0.160 | 0.019 |
| 13.8 | 7.08 | 3.43 | 0.12 | 0.17 | 4.14 | 0.77 | 0.144 | 0.020 |
| 14.6 | 7.09 | 3.88 | 0.13 | 0.20 | 4.36 | 0.81 | 0.124 | 0.019 |
| 11.5 | 6.37 | 5.64 | 0.16 | 0.14 | 3.89 | 0.69 | 0.087 | 0.009 |
| 13.8 | 7.06 | 5.34 | 0.20 | 0.20 | 4.31 | 0.71 | 0.101 | 0.021 |
| 10.9 | 6.26 | 3.47 | 0.17 | 0.22 | 3.40 | 0.66 | 0.109 | 0.010 |
| 10.1 | 4.26 | 4.26 | 0.20 | 0.18 | 3.32 | 0.59 | 0.149 | 0.017 |
| 11.6 | 5.78 | 5.91 | 0.18 | 0.16 | 3.70 | 0.65 | 0.082 | 0.015 |
| 11.1 | 5.49 | 4.52 | 0.29 | 0.30 | 4.03 | 0.63 | 0.080 | 0.016 |
| 13.4 | 6.92 | 7.23 | 0.28 | 0.20 | 4.54 | 0.69 | 0.163 | 0.017 |
| 12.4 | 6.13 | 5.69 | 0.22 | 0.54 | 4.65 | 0.70 | 0.159 | 0.015 |
| 13.1 | 6.64 | 5.51 | 0.31 | 0.24 | 4.89 | 0.72 | 0.094 | 0.017 |
| 11.6 | 5.39 | 3.77 | 0.29 | 0.06 | 4.51 | 0.56 | 0.110 | 0.015 |
| 11.8 | 5.44 | 3.94 | 0.30 | 0.04 | 4.64 | 0.59 | 0.085 | 0.013 |
| 18.3 | 1.28 | 0.67 | 0.03 | 0.03 | 1.96 | 0.65 | 0.001 | 0.014 |
| 15.8 | 2.39 | 0.63 | 0.01 | 0.04 | 1.94 | 1.29 | 0.001 | 0.014 |
| 18.0 | 1.50 | 0.67 | 0.01 | 0.06 | 2.11 | 0.92 | 0.001 | 0.016 |

**Table 15.3**: `pottery` data (continued).

| Al2O3 | Fe2O3 | MgO | CaO | Na2O | K2O | TiO2 | MnO | BaO |
|---|---|---|---|---|---|---|---|---|
| 18.0 | 1.88 | 0.68 | 0.01 | 0.04 | 2.00 | 1.11 | 0.006 | 0.022 |
| 20.8 | 1.51 | 0.72 | 0.07 | 0.10 | 2.37 | 1.26 | 0.002 | 0.016 |
| 17.7 | 1.12 | 0.56 | 0.06 | 0.06 | 2.06 | 0.79 | 0.001 | 0.013 |
| 18.3 | 1.14 | 0.67 | 0.06 | 0.05 | 2.11 | 0.89 | 0.006 | 0.019 |
| 16.7 | 0.92 | 0.53 | 0.01 | 0.05 | 1.76 | 0.91 | 0.004 | 0.013 |
| 14.8 | 2.74 | 0.67 | 0.03 | 0.05 | 2.15 | 1.34 | 0.003 | 0.015 |
| 19.1 | 1.64 | 0.60 | 0.10 | 0.03 | 1.75 | 1.04 | 0.007 | 0.018 |

*Source*: Tubb, A., et al., *Archaeometry*, 22, 153–171, 1980. With permission.

Ex. 15.2 Construct a three-dimensional drop-line scatterplot of the planets data in which the points are labelled with a suitable cluster label.

Ex. 15.3 Write an R function to fit a mixture of $k$ normal densities to a data set using maximum likelihood.

Ex. 15.4 A class of cluster analysis techniques not considered in this chapter contains the *agglomerative hierarchical techniques*. These are described in detail in Everitt et al. (2001). Members of the class can be implemented in R using the functions, `hclust` and `dist`, and the resulting *dendrograms* can be plotted using the function, `plclust`. Look at the relevant help pages to see how to use these functions and then apply *complete linkage* and *average linkage* to the `planets` data. Compare the results with those given in the text.

Ex. 15.5 Write a general R function that will display a particular partition from the $k$-means cluster method on both a scatterplot matrix of the original data and a scatterplot or scatterplot matrix of a selected number of principal components of the data.

# Bibliography


Agresti, A. (1996), *An Introduction to Categorical Data Analysis*, New York, USA: John Wiley & Sons.

Agresti, A. (2002), *Categorical Data Analysis*, Hoboken, New Jersey, USA: John Wiley & Sons, 2nd edition.

Aitkin, M. (1978), "The analysis of unbalanced cross-classifications," *Journal of the Royal Statistical Society, Series A*, 141, 195–223, with discussion.

Alon, U., Barkai, N., Notternam, D. A., Gish, K., Ybarra, S., Mack, D., and Levine, A. J. (1999), "Broad patterns of gene expressions revealed by clustering analysis of tumour and normal colon tissues probed by oligonucleotide arrays," *Cell Biology*, 99, 6754–6760.

Ambler, G. and Benner, A. (2005), *mfp: Multivariable Fractional Polynomials*, URL `http://CRAN.R-project.org`, R package version 1.3.1.

Aspirin Myocardial Infarction Study Research Group (1980), "A randomized, controlled trial of aspirin in persons recovered from myocardial infarction," *Journal of the American Medical Association*, 243, 661–669.

Bailey, K. R. (1987), "Inter-study differences: how should they influence the interpretation of results?" *Statistics in Medicine*, 6, 351–360.

Banfield, J. D. and Raftery, A. E. (1993), "Model-based gaussian and non-gaussian clustering," *Biometrics*, 49, 803–821.

Barlow, R. E., Bartholomew, D. J., Bremner, J. M., and Brunk, H. D. (1972), *Statistical Inference under Order Restrictions*, New York, USA: John Wiley & Sons.

Bates, D. (2005), "Fitting linear mixed models in R," *R News*, 5, 27–30, URL `http://CRAN.R-project.org/doc/Rnews/`.

Bates, D. and Sarkar, D. (2005), *lme4: Linear Mixed-Effects Models Using S4 Classes*, URL `http://CRAN.R-project.org`, R package version 0.98-1.

Beck, A., Steer, R., and Brown, G. (1996), *BDI-II Manual*, The Psychological Corporation, San Antonio, 2nd edition.

Becker, R. A., Chambers, J. M., and Wilks, A. R. (1988), *The New S Language*, London, UK: Chapman & Hall.

Breddin, K., Loew, D., Lechner, K., Überla, K., and Walter, E. (1979), "Secondary prevention of myocardial infarction. Comparison of acetylsalicylic acid, phenprocoumon and placebo. A multicenter two-year prospective study," *Thrombosis and Haemostasis*, 41, 225–236.

Breiman, L. (1996), "Bagging predictors," *Machine Learning*, 24, 123–140.

Breiman, L. (2001a), "Random forests," *Machine Learning*, 45, 5–32.

Breiman, L. (2001b), "Statistical modeling: The two cultures," *Statistical Science*, 16, 199–231, with discussion.

Breiman, L., Cutler, A., Liaw, A., and Wiener, M. (2005), *randomForest: Breiman and Cutler's Random Forests for Classification and Regression*, URL `http://stat-www.berkeley.edu/users/breiman/RandomForests`, R package version 4.5-15.

Breiman, L., Friedman, J. H., Olshen, R. A., and Stone, C. J. (1984), *Classification and Regression Trees*, California, USA: Wadsworth.

Bühlmann, P. (2004), "Bagging, boosting and ensemble methods," in *Handbook of Computational Statistics*, eds. J. E. Gentle, W. Härdle, and Y. Mori, Springer, Berlin, Heidelberg, pp. 877–907.

Canty, A. and Ripley, B. D. (2005), *boot: Bootstrap R (S-PLUS) Functions (Canty)*, URL `http://CRAN.R-project.org`, R package version 1.2-24.

Carey, V. J., Lumley, T., and Ripley, B. D. (2002), *gee: Generalized Estimation Equation Solver*, URL `http://CRAN.R-project.org`, R package version 4.13-10.

Carlin, J. B., Ryan, L. M., Harvey, E. A., and Holmes, L. B. (2000), "Anticonvulsant teratogenesis 4: Inter-rater agreement in assessing minor physical features related to anticonvulsant therapy," *Teratology*, 62, 406–412.

Carpenter, J., Pocock, S., and Lamm, C. J. (2002), "Coping with missing data in clinical trials: A model-based approach applied to asthma trials," *Statistics in Medicine*, 21, 1043–1066.

Chalmers, T. C. and Lau, J. (1993), "Meta-analytic stimulus for changes in clinical trials," *Statistical Methods in Medical Research*, 2, 161–172.

Chambers, J. M. (1998), *Programming with Data*, New York, USA: Springer.

Chambers, J. M. and Hastie, T. J. (1992), *Statistical Models in S*, London, UK: Chapman & Hall.

Colditz, G. A., Brewer, T. F., Berkey, C. S., Wilson, M. E., Burdick, E., Fineberg, H. V., and Mosteller, F. (1994), "Efficacy of BCG vaccine in the prevention of tuberculosis. Meta-analysis of the published literature," *Journal of the American Medical Association*, 271, 698–702.

Collett, D. (2003), *Modelling Binary Data*, London, UK: Chapman & Hall/CRC, 2nd edition.

Collett, D. and Jemain, A. A. (1985), "Residuals, outliers and influential observations in regression analysis," *Sains Malaysiana*, 4, 493–511.

Cook, R. D. and Weisberg, S. (1982), *Residuals and Influence in Regression*, London, UK: Chapman & Hall/CRC.

Cook, R. J. (1998), "Generalized linear model," in *Encyclopedia of Biostatistics*, eds. P. Armitage and T. Colton, Chichester, UK: John Wiley & Sons.

Corbet, G. B., Cummins, J., Hedges, S. R., and Krzanowski, W. J. (1970), "The taxonomic structure of British water voles, genus *Arvicola*," *Journal of Zoology*, 61, 301–316.

Coronary Drug Project Group (1976), "Asprin in coronary heart disease," *Journal of Chronic Diseases*, 29, 625–642.

Cox, D. R. (1972), "Regression models and life-tables," *Journal of the Royal Statistical Society, Series B*, 34, 187–202, with discussion.

Dalgaard, P. (2002), *Introductory Statistics with R*, New York: Springer.

Davis, C. S. (1991), "Semi-parametric and non-parametric methods for the analysis of repeated measurements with applications to clinical trials," *Statistics in Medicine*, 10, 1959–1980.

Davis, C. S. (2002), *Statistical Methods for the Analysis of Repeated Measurements*, New York, USA: Springer.

DeMets, D. L. (1987), "Methods for combining randomized clinical trials: strengths and limitations," *Statistics in Medicine*, 6, 341–350.

Dempster, A. P., Laird, N. M., and Rubin, D. B. (1977), "Maximum likelihood from incomplete data via the EM algorithm (C/R: p22-37)," *Journal of the Royal Statistical Society, Series B*, 39, 1–22.

DerSimonian, R. and Laird, N. (1986), "Meta-analysis in clinical trials," *Controlled Clinical Trials*, 7, 177–188.

Diggle, P. J. (1998), "Dealing with missing values in longitudinal studies," in *Statistical Analysis of Medical Data*, eds. B. S. Everitt and G. Dunn, London, UK: Arnold.

Diggle, P. J., Heagerty, P. J., Liang, K. Y., and Zeger, S. L. (2003), *Analysis of Longitudinal Data*, Oxford, UK: Oxford University Press.

Diggle, P. J. and Kenward, M. G. (1994), "Informative dropout in longitudinal data analysis," *Journal of the Royal Statistical Society, Series C*, 43, 49–93.

Dolnicar, S. and Leisch, F. (2003), "Winter tourist segments in Austria: Identifying stable vacation styles using bagged clustering techniques," *Journal of Travel Research*, 41, 281–292.

Dubes, R. and Jain, A. K. (1979), "Validity studies in clustering methodologies," *Pattern Recognition*, 8, 247–260.

Duval, S. and Tweedie, R. L. (2000), "A nonparametric 'trim and fill' method of accounting for publication bias in meta-analysis," *Journal of the American Statistical Association*, 95, 89–98.

Easterbrook, P. J., Berlin, J. A., Gopalan, R., and Matthews, D. R. (1991), "Publication bias in research," *Lancet*, 337, 867–872.

Edgington, E. S. (1987), *Randomization Tests*, New York, USA: Marcel Dekker.

Efron, B. and Tibshirani, R. J. (1993), *An Introduction to the Bootstrap*, London, UK: Chapman & Hall/CRC.

Elwood, P. C., Cochrane, A. L., Burr, M. L., Sweetman, P. M., Williams, G., Welsby, E., Hughes, S. J., and Renton, R. (1974), "A randomized controlled trial of acetyl salicilic acid in the secondary prevention of mortality from myocardial infarction," *British Medical Journal*, 1, 436–440.

Elwood, P. C. and Sweetman, P. M. (1979), "Asprin and secondary mortality after myocardial infarction," *Lancet*, 2, 1313–1315.

Everitt, B. S. (1992), *The Analysis of Contingency Tables*, London, UK: Chapman & Hall/CRC, 2nd edition.

Everitt, B. S. (1996), *Making Sense of Statistics in Psychology: A Second-Level Course*, Oxford, UK: Oxford University Press.

Everitt, B. S. (2001), *Statistics for Psychologists*, Mahwah, New Jersey, USA: Lawrence Erlbaum.

Everitt, B. S. (2002a), *Cambridge Dictionary of Statistics in the Medical Sciences*, Cambridge, UK: Cambridge University Press.

Everitt, B. S. (2002b), *Modern Medical Statistics*, London, UK: Arnold.

Everitt, B. S. and Bullmore, E. T. (1999), "Mixture model mapping of brain activation in functional magnetic resonance images," *Human Brain Mapping*, 7, 1–14.

Everitt, B. S. and Dunn, G. (2001), *Applied Multivariate Data Analysis*, London, UK: Arnold, 2nd edition.

Everitt, B. S., Landau, S., and Leese, M. (2001), *Cluster Analysis*, London, UK: Arnold, 4th edition.

Everitt, B. S. and Pickles, A. (2000), *Statistical Aspects of the Design and Analysis of Clinical Trials*, London, UK: Imperial College Press.

Everitt, B. S. and Rabe-Hesketh, S. (1997), *The Analysis of Proximity Data*, London, UK: Arnold.

Everitt, B. S. and Rabe-Hesketh, S. (2001), *Analysing Medical Data Using S-Plus*, New York, USA: Springer.

Fisher, R. A. (1935), *The Design of Experiments*, Edinburgh, UK: Oliver and Boyd.

Fleiss, J. L. (1993), "The statistical basis of meta-analysis," *Statistical Methods in Medical Research*, 2, 121–145.

Fraley, C. and Raftery, A. E. (2002), "Model-based clustering, discriminant analysis, and density estimation," *Journal of the American Statistical Association*, 97, 611–631.

Fraley, C., Raftery, A. E., and Wehrens, R. (2005), *mclust: Model-based Cluster Analysis*, URL `http://www.stat.washington.edu/mclust`, R package version 2.1-11.

Fraley, G. and Raftery, A. E. (1999), "MCLUST: Software for model-based cluster analysis," *Journal of Classification*, 16, 297–306.

Freeman, G. H. and Halton, J. H. (1951), "Note on an exact treatment of contingency, goodness of fit and other problems of significance," *Biometrika*, 38, 141–149.

Friedman, H. P. and Rubin, J. (1967), "On some invariant criteria for grouping data," *Journal of the American Statistical Association*, 62, 1159–1178.

Gabriel, K. R. (1971), "The biplot graphical display of matrices with application to principal component analysis," *Biometrika*, 58, 453–467.

Garczarek, U. M. and Weihs, C. (2003), "Standardizing the comparison of partitions," *Computational Statistics*, 18, 143–162.

Gentleman, R. (2005), "Reproducible research: A bioinformatics case study," *Statistical Applications in Genetics and Molecular Biology*, 4, URL `http://www.bepress.com/sagmb/vol4/iss1/art2`, article 2.

Giardiello, F. M., Hamilton, S. R., Krush, A. J., Piantadosi, S., Hylind, L. M., Celano, P., Booker, S. V., Robinson, C. R., and Offerhaus, G. J. A. (1993), "Treatment of colonic and rectal adenomas with sulindac in familial adenomatous polyposis," *New England Journal of Medicine*, 328, 1313–1316.

Gordon, A. (1999), *Classification*, Boca Raton, Florida, USA: Chapman & Hall/CRC, 2nd edition.

Gower, J. C. and Ross, G. J. S. (1969), "Minimum spanning trees and single linkage cluster analysis," *Applied Statistics*, 18, 54–64.

Grana, C., Chinol, M., Robertson, C., Mazzetta, C., Bartolomei, M., Cicco, C. D., Fiorenza, M., Gatti, M., Caliceti, P., and Paganelli1, G. (2002), "Pretargeted adjuvant radioimmunotherapy with Yttrium-90-biotin in malignant glioma patients: A pilot study," *British Journal of Cancer*, 86, 207–212.

Greenwald, A. G. (1975), "Consequences of prejudice against the null hypothesis," *Psychological Bulletin*, 82, 1–20.

Greenwood, M. and Yule, G. U. (1920), "An inquiry into the nature of frequency distribution of multiple happenings with particular reference of multiple attacks of disease or of repeated accidents," *Journal of the Royal Statistical Society*, 83, 255–279.

Haberman, S. J. (1973), "The analysis of residuals in cross-classified tables," *Biometrics*, 29, 205–220.

Hand, D. J., Daly, F., Lunn, A. D., McConway, K. J., and Ostrowski, E. (1994), *A Handbook of Small Datasets*, London, UK: Chapman & Hall/CRC.

Harrison, D. and Rubinfeld, D. L. (1978), "Hedonic prices and the demand for clean air," *Journal of Environmental Economics & Management*, 5, 81–102.

Hartigan, J. A. (1975), *Clustering Algorithms*, New York, USA: John Wiley & Sons.

Heitjan, D. F. (1997), "Annotation: What can be done about missing data? Approaches to imputation," *American Journal of Public Health*, 87, 548–550.

Hochberg, Y. and Tamhane, A. C. (1987), *Multiple Comparison Procedures*, New York, USA: John Wiley & Sons.

Hothorn, T., Hornik, K., van de Wiel, M., and Zeileis, A. (2005a), *coin: Conditional Inference Procedures in a Permutation Test Framework*, URL `http://CRAN.R-project.org`, R package version 0.4-2.

Hothorn, T., Hornik, K., van de Wiel, M. A., and Zeileis, A. (2006a), "A Lego system for conditional inference," *The American Statistician*, 60, 257–263.

Hothorn, T., Hornik, K., and Zeileis, A. (2005b), *party: A Laboratory for Recursive Part(y)itioning*, URL `http://CRAN.R-project.org`, R package version 0.3-1.

Hothorn, T., Hornik, K., and Zeileis, A. (2006b), "Unbiased recursive partitioning: A conditional inference framework," *Journal of Computational and Graphical Statistics*, 15, 651–674.

ISIS-2 (Second International Study of Infarct Survival) Collaborative Group (1988), "Randomised trial of intravenous streptokinase, oral aspirin, both, or neither among 17,187 cases of suspected acute myocardial infarction: ISIS-2," *Lancet*, 13, 349–360.

Kalbfleisch, J. D. and Prentice, R. L. (1980), *The Statistical Analysis of Failure Time Data*, New York, USA: John Wiley & Sons.

Kaplan, E. L. and Meier, P. (1958), "Nonparametric estimation from incomplete observations," *Journal of the American Statistical Association*, 53, 457–481.

Kaufman, L. and Rousseeuw, P. J. (1990), *Finding Groups in Data: An Introduction to Cluster Analysis*, New York, USA: John Wiley & Sons.

Kraepelin, E. (1919), *Dementia Praecox and Paraphrenia*, Edinburgh, UK: Livingstone.

Kruskal, J. B. (1964a), "Multidimensional scaling by optimizing goodness-of-fit to a nonmetric hypothesis," *Psychometrika*, 29, 1–27.

Kruskal, J. B. (1964b), "Nonmetric multidimensional scaling: A numerical method," *Psychometrika*, 29, 115–129.

Lanza, F. L. (1987), "A double-blind study of prophylactic effect of misoprostol on lesions of gastric and duodenal mucosa induced by oral administration of tolmetin in healthy subjects," *British Journal of Clinical Practice*, 40, 91–101.

Lanza, F. L., Aspinall, R. L., Swabb, E. A., Davis, R. E., Rack, M. F., and Rubin, A. (1988a), "Double-blind, placebo-controlled endoscopic comparison of the mucosal protective effects of misoprostol versus cimetidine on tolmetin-induced mucosal injury to the stomach and duodenum," *Gastroenterology*, 95, 289–294.

Lanza, F. L., Fakouhi, D., Rubin, A., Davis, R. E., Rack, M. F., Nissen, C., and Geis, S. (1989), "A double-blind placebo-controlled comparison of the efficacy and safety of 50, 100, and 200 micrograms of misoprostol QID in the prevention of Ibuprofen-induced gastric and duodenal mucosal lesions and symptoms," *American Journal of Gastroenterology*, 84, 633–636.

Lanza, F. L., Peace, K., Gustitus, L., Rack, M. F., and Dickson, B. (1988b), "A blinded endoscopic comparative study of misoprostol versus sucralfate and placebo in the prevention of aspirin-induced gastric and duodenal ulceration," *American Journal of Gastroenterology*, 83, 143–146.

Leisch, F. (2002a), "Sweave: Dynamic generation of statistical reports using literate data analysis," in *Compstat 2002 — Proceedings in Computational Statistics*, eds. W. Härdle and B. Rönz, Physica Verlag, Heidelberg, pp. 575–580, ISBN 3-7908-1517-9.

Leisch, F. (2002b), "Sweave, Part I: Mixing R and LaTeX," *R News*, 2, 28–31, URL `http://CRAN.R-project.org/doc/Rnews/`.

Leisch, F. (2003), "Sweave, Part II: Package vignettes," *R News*, 3, 21–24, URL `http://CRAN.R-project.org/doc/Rnews/`.

Leisch, F. (2004), "FlexMix: A general framework for finite mixture models and latent class regression in R," *Journal of Statistical Software*, 11, URL `http://www.jstatsoft.org/v11/i08/`.

Leisch, F. and Dimitriadou, E. (2005), *mlbench: Machine Learning Benchmark Problems (Original data sets from various sources)*, URL `http://CRAN.R-project.org`, R package version 1.1-0.

Leisch, F. and Rossini, A. J. (2003), "Reproducible statistical research," *Chance*, 16, 46–50.

Liang, K. and Zeger, S. L. (1986), "Longitudinal data analysis using generalized linear models," *Biometrika*, 73, 13–22.

Ligges, U. and Mächler, M. (2003), "Scatterplot3d – an R package for visualizing multivariate data," *Journal of Statistical Software*, 8, 1–20, URL `http://www.jstatsoft.org`.

Longford, N. T. (1993), *Random Coefficient Models*, Oxford, UK: Oxford University Press.

Lumley, T. (2005), *rmeta: Meta-Analysis*, URL `http://CRAN.R-project.org`, R package version 2.12.

Lumley, T. and Miller, A. (2005), *leaps: Regression Subset Selection*, URL `http://CRAN.R-project.org`, R package version 2.7.

Mann, L. (1981), "The baiting crowd in episodes of threatened suicide," *Journal of Personality and Social Psychology*, 41, 703–709.

Mardia, K. V., Kent, J. T., and Bibby, J. M. (1979), *Multivariate Analysis*, London, UK: Academic Press.

Mardin, C. Y., Hothorn, T., Peters, A., Jünemann, A. G., Nguyen, N. X., and Lausen, B. (2003), "New glaucoma classification method based on standard HRT parameters by bagging classification trees," *Journal of Glaucoma*, 12, 340–346.

Marriott, F. H. C. (1982), "Optimization methods of cluster analysis," *Biometrika*, 69, 417–421.

Mayor, M. and Frei, P. (2003), *New Worlds in the Cosmos: The Discovery of Exoplanets*, Cambridge, UK: Cambridge University Press.

Mayor, M. and Queloz, D. (1995), "A Jupiter-mass companion to a solar-type star," *Nature*, 378, 355.

McCullagh, P. and Nelder, J. A. (1989), *Generalized Linear Models*, London, UK: Chapman & Hall/CRC.

McLachlan, G. and Peel, D. (2000), *Finite Mixture Models*, New York, USA: John Wiley & Sons.

Mehta, C. R. and Patel, N. R. (2003), *StatXact-6: Statistical Software for Exact Nonparametric Inference*, Cytel Software Corporation, Cambridge, MA, USA.

Meyer, D., Zeileis, A., Karatzoglou, A., and Hornik, K. (2005), *vcd: Visualizing Categorical Data*, URL `http://CRAN.R-project.org`, R package version 0.9-7.

Miller, A. (2002), *Subset Selection in Regression*, New York, USA: Chapman & Hall, 2nd edition.

Morrison, D. F. (2005), "Multivariate analysis of variance," in *Encyclopedia of Biostatistics*, eds. P. Armitage and T. Colton, Chichester, UK: John Wiley & Sons, 2nd edition.

Murray, G. D. and Findlay, J. G. (1988), "Correcting for bias caused by dropouts in hypertension trials," *Statistics in Medicine*, 7, 941–946.

Murrell, P. (2005), *R Graphics*, Boca Raton, Florida, USA: Chapman & Hall/CRC.

Murthy, S. K. (1998), "Automatic construction of decision trees from data: A multi-disciplinary survey," *Data Mining and Knowledge Discovery*, 2, 345–389.

Nelder, J. A. (1977), "A reformulation of linear models," *Journal of the Royal Statistical Society, Series A*, 140, 48–76, with commentary.

Nelder, J. A. and Wedderburn, R. W. M. (1972), "Generalized linear models," *Journal of the Royal Statistical Society, Series A*, 135, 370–384.

Oakes, M. (1993), "The logic and role of meta-analysis in clinical research," *Statistical Methods in Medical Research*, 2, 147–160.

Paradis, E., Strimmer, K., Claude, J., Jobb, G., Opgen-Rhein, R., Dutheil, J., Noel, Y., and Bolker, B. (2005), *ape: Analyses of Phylogenetics and Evolution*, URL `http://CRAN.R-project.org`, R package version 1.8.

Pearson, K. (1894), "Contributions to the mathematical theory of evolution," *Philosophical Transactions A*, 185, 71–110.

Persantine-Aspirin Reinfarction Study Research Group (1980), "Persantine and Aspirin in coronary heart disease," *Circulation*, 62, 449–461.

Pesarin, F. (2001), *Multivariate Permutation Tests: With Applications to Biostatistics*, Chichester, UK: John Wiley & Sons.

Peters, A., Hothorn, T., and Lausen, B. (2002), "ipred: Improved predictors," *R News*, 2, 33–36, URL `http://CRAN.R-project.org/doc/Rnews/`, ISSN 1609-3631.

Petitti, D. B. (2000), *Meta-Analysis, Decision Analysis and Cost-Effectiveness Analysis*, New York, USA: Oxford University Press.

Piantadosi, S. (1997), *Clinical Trials: A Methodologic Perspective*, New York, USA: John Wiley & Sons.

Pinheiro, J. C. and Bates, D. M. (2000), *Mixed-Effects Models in S and S-PLUS*, New York, USA: Springer.

Pitman, E. J. G. (1937), "Significance tests which may be applied to samples from any populations," *Biometrika*, 29, 322–335.

Postman, M., Huchra, J. P., and Geller, M. J. (1986), "Probes of large-scale structures in the corona borealis region," *Astrophysical Journal*, 92, 1238–1247.

Prim, R. C. (1957), "Shortest connection networks and some generalizations," *Bell System Technical Journal*, 36, 1389–1401.

Proudfoot, J., Goldberg, D., Mann, A., Everitt, B. S., Marks, I., and Gray, J. A. (2003), "Computerized, interactive, multimedia cognitive-behavioural program for anxiety and depression in general practice," *Psychological Medicine*, 33, 217–227.

Quine, S. (1975), *Achievement Orientation of Aboriginal and White Adolescents*, Doctoral Dissertation, Australian National University, Canberra, Australia.

R Development Core Team (2005a), *An Introduction to R*, R Foundation for Statistical Computing, Vienna, Austria, URL `http://www.R-project.org`, ISBN 3-900051-12-7.

R Development Core Team (2005b), *R: A Language and Environment for Statistical Computing*, R Foundation for Statistical Computing, Vienna, Austria, URL `http://www.R-project.org`, ISBN 3-900051-07-0.

R Development Core Team (2005c), *R Data Import/Export*, R Foundation for Statistical Computing, Vienna, Austria, URL `http://www.R-project.org`, ISBN 3-900051-10-0.

R Development Core Team (2005d), *R Installation and Administration*, R Foundation for Statistical Computing, Vienna, Austria, URL `http://www.R-project.org`, ISBN 3-900051-09-7.

R Development Core Team (2005e), *Writing R Extensions*, R Foundation for Statistical Computing, Vienna, Austria, URL `http://www.R-project.org`, ISBN 3-900051-11-9.

Ripley, B. D. (1996), *Pattern Recognition and Neural Networks*, Cambridge, UK: Cambridge University Press, URL http://www.stats.ox.ac.uk/pub/PRNN/.

Roeder, K. (1990), "Density estimation with confidence sets exemplified by superclusters and voids in galaxies," *Journal of the American Statistical Association*, 85, 617–624.

Romesburg, H. C. (1984), *Cluster Analysis for Researchers*, Belmont, CA: Lifetime Learning Publications.

Rubin, D. (1976), "Inference and missing data," *Biometrika*, 63, 581–592.

Sauerbrei, W. and Royston, P. (1999), "Building multivariable prognostic and diagnostic models: Transformation of the predictors by using fractional polynomials," *Journal of the Royal Statistical Society, Series A*, 162, 71–94.

Schumacher, M., Basert, G., Bojar, H., Hübner, K., Olschewski, M., Sauerbrei, W., Schmoor, C., Beyerle, C., Neumann, R. L. A., and Rauschecker, H. F. for the German Breast Cancer Study Group (1994), "Randomized $2 \times 2$ trial evaluating hormonal treatment and the duration of chemotherapy in node-positive breast cancer patients," *Journal of Clinical Oncology*, 12, 2086–2093.

Scott, A. J. and Symons, M. J. (1971), "Clustering methods based on likelihood ratio criteria," *Biometrics*, 27, 387–398.

Scott, D. W. (1992), *Multivariate Density Estimation*, New York, USA: John Wiley & Sons.

Searle, S. R. (1971), *Linear Models*, New York, USA: John Wiley & Sons.

Seeber, G. U. H. (1998), "Poisson regression," in *Encyclopedia of Biostatistics*, eds. P. Armitage and T. Colton, Chichester, UK: John Wiley & Sons.

Shepard, R. N. (1962a), "The analysis of proximities: Multidimensional scaling with unknown distance function Part I," *Psychometrika*, 27, 125–140.

Shepard, R. N. (1962b), "The analysis of proximities: Multidimensional scaling with unknown distance function Part II," *Psychometrika*, 27, 219–246.

Sibson, R. (1979), "Studies in the robustness of multidimensional scaling. Perturbational analysis of classical scaling," *Journal of the Royal Statistical Society, Series B*, 41, 217–229.

Silagy, C. (2003), "Nicotine replacement therapy for smoking cessation (Cochrane Review)," in *The Cochrane Library*, John Wiley & Sons, issue 4.

Silverman, B. (1986), *Density Estimation*, London, UK: Chapman & Hall/CRC.

Simonoff, J. S. (1996), *Smoothing Methods in Statistics*, New York, USA: Springer.

Smith, M. L. (1980), "Publication bias and meta-analysis," *Evaluating Education*, 4, 22–93.

Sterlin, T. D. (1959), "Publication decisions and their possible effects on inferences drawn from tests of significance-or vice versa," *Journal of the American Statistical Association*, 54, 30–34.

Stevens, J. (2001), *Applied Multivariate Statistics for the Social Sciences*, Mahwah, New Jersey, USA: Lawrence Erlbaum, 4th edition.

Sutton, A. J. and Abrams, K. R. (2001), "Bayesian methods in meta-analysis and evidence synthesis," *Statistical Methods in Medical Research*, 10, 277–303.

Sutton, A. J., Abrams, K. R., Jones, D. R., and Sheldon, T. A. (2000), *Methods for Meta-Analysis in Medical Research*, Chichester, UK: John Wiley & Sons.

Thall, P. F. and Vail, S. C. (1990), "Some covariance models for longitudinal count data with overdispersion," *Biometrics*, 46, 657–671.

Therneau, T. M., Atkinson, B., and Ripley, B. D. (2005), *rpart: Recursive Partitioning*, URL `http://www.mayo.edu/hsr/Sfunc.html`, R package version 3.1-27.

Therneau, T. M. and Atkinson, E. J. (1997), "An introduction to recursive partitioning using the rpart routine," Technical Report 61, Section of Biostatistics, Mayo Clinic, Rochester, USA, URL `http://www.mayo.edu/hsr/techrpt/61.pdf`.

Therneau, T. M. and Grambsch, P. M. (2000), *Modeling Survival Data: Extending the Cox Model*, New York, USA: Springer.

Therneau, T. M. and Lumley, T. (2005), *survival: Survival Analysis, including Penalised Likelihood*, URL `http://CRAN.R-project.org`, R package version 2.20.

Timm, N. H. (2002), *Applied Multivariate Analysis*, New York, USA: Springer.

Tubb, A., Parker, N. J., and Nickless, G. (1980), "The analysis of Romano-British pottery by atomic absorption spectrophotometry," *Archaeometry*, 22, 153–171.

Tukey, J. W. (1953), "The problem of multiple comparisons (unpublished manuscript)," in *The Collected Works of John W. Tukey VIII. Multiple Comparisons: 1948-1983*, New York, USA: Chapman & Hall.

Vanisma, F. and De Greve, J. P. (1972), "Close binary systems before and after mass transfer," *Astrophysics and Space Science*, 87, 377–401.

Venables, W. N. and Ripley, B. D. (2002), *Modern Applied Statistics with S*, Springer, 4th edition, URL `http://www.stats.ox.ac.uk/pub/MASS4/`, ISBN 0-387-95457-0.

Wand, M. P. and Jones, M. C. (1995), *Kernel Smoothing*, London, UK: Chapman & Hall/CRC.

Wand, M. P. and Ripley, B. D. (2005), *KernSmooth: Functions for Kernel Smoothing for Wand & Jones (1995)*, URL `http://www.maths.unsw.edu.au/~wand`, R package version 2.22-16.

Whitehead, A. and Jones, N. M. B. (1994), "A meta-analysis of clinical trials involving different classifications of response into ordered categories," *Statistics in Medicine*, 13, 2503–2515.

Wilkinson, L. (1992), "Graphical displays," *Statistical Methods in Medical Research*, 1, 3–25.

Woodley, W. L., Simpson, J., Biondini, R., and Berkeley, J. (1977), "Rainfall results 1970-75: Florida area cumulus experiment," *Science*, 195, 735–742.

Young, G. and Householder, A. S. (1938), "Discussion of a set of points in terms of their mutual distances," *Psychometrika*, 3, 19–22.

Zeger, S. L. and Liang, K. Y. (1986), "Longitudinal data analysis for discrete and continuous outcomes," *Biometrics*, 42, 121–130.

Zeileis, A. (2004), "Econometric computing with HC and HAC covariance matrix estimators," *Journal of Statistical Software*, 11, 1–17, URL `http://jstatsoft.org/v11/i10/`.

# Index